WOMEN

AND

NUMBERS

by Teri Perl

Lives of
Women Mathematicians
plus Discovery Activities

WIDE WORLD PUBLISHING/TETRA

Illustrations by Analee Nunan.

Wide World Publishing/Tetra
P.O. Box 476
San Carlos, CA 94070

Printed in the United States of America.

2nd Printing November 1994

ISBN: 0-933174-87-X

Library of Congress Cataloging-in-Publication Data

Perl, Teri .
    Women and numbers : lives of women mathematicians plus discovery
activities / Teri Perl .
        p.   cm.
    Summary : Presents biographies of women from the nineteenth and
twentieth centuries who pursued their interests in mathematics . Each
chapter includes different mathematical activities.
        ISBN 0 - 933174 -87 -X
        1. Women mathematicians - - Biography - - Juvenile literature.
[1. Women mathematicians . 2. Mathematicians. 3. Mathematical
recreations.]   I. Title.
QA28 . P48   1993
510 ' . 92 ' 2 - - dc20
        [B]   93 - 10535
                                                              CIP
                                                              AC

ACKNOWLEDGEMENTS

First we wish to thank all the people who helped make the earlier edition of the book possible, in particular the Women's Educational Equity Program of the United States Department of Education. WEEA was the agency that funded the project in 1980. As before, we extend special thanks to all the contemporary subjects of the biographies for their enthusiasm and patience during the interview, writing, and re-writing process, and for sharing so freely of their life experiences.

In the earlier version we specifically acknowledged the participants and leaders at EQUALS seminars who evaluated the work as it was being written. We wish to do so again. Over the years EQUALS has grown into an important program for raising educators' consciousness about equity issues. One of the major enhancements to this current edition is a chapter that looks at EQUALS and several staff members who are important in its growing success.

To Analee Nunan, my longest California friend, many thanks for sharing your skills as an artist to enrich the book immeasurably.

I wish also to thank the wonderful women at the National Women's History Project for keeping the first edition of this work in print over the years. And finally, to my current publishers, thank you for helping make this book available to a new generation of young readers.

PREFACE

"Dream big and keep your options wide open"

This is the last decade of the twentieth century, and most women expect to work the greater part of their adult lives. We want to share the stories of a group of women whose work has been very important to them. Some of these women lived long before you were born. Like you, they are from different backgrounds.

To all of them, their work was rewarding in many ways. They worked on interesting problems. They were proud of the results of their work. Among the contemporary women, the fruits of their work allow them independence. Women such as these, who find their work rewarding and challenging, are the lucky ones.

In the section called *Expanding Your Horizons* you will read about a conference to inspire young women to join these lucky ones. It is a true story, as are all the other stories in this book.

Young women growing up now can become whatever their talents and dreams allow. Become one of the lucky ones. Dream big and keep your options wide open.

TABLE OF CONTENTS

BIOGRAPHIES & ACTIVITIES

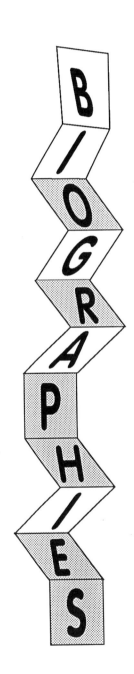

MARY SOMERVILLE
1780-1872

**"It was commonly believed that
girls and women could go mad if
they exercised their minds too much."**

Mary Somerville was born in Burntisland, a small village on the northern seacoast of Scotland. As a child she loved to roam outdoors. She had only to step through an opening at the end of her garden and she was at the seashore.

"When the tide was out," she wrote, "I spent hours on the sands, looking at the star-fish and sea-urchins, or watching the children digging for sand-eels, cockels, and the spouting razorfish. I knew the eggs of many birds, and made a collection of them. I never robbed a nest, but bought strings of eggs, which were sold by boys. I also watched the crabs, live shells, jellyfish, and various marine animals, all of which were objects of curiosity and amusement to me in my lonely life."

Mary's father was an Admiral in the English Navy. When Mary was about ten years old he came home after a long voyage. He noticed how wild Mary had become from her constant outdoor explorations and insisted she be sent off to a boarding school. As a result, Mary spent a year at Miss Primrose's boarding school.

Miss Primrose's School

Away from the wild Scottish coast, Mary was desperately unhappy.

At school she learned reading, writing, French and English grammar. The method of teaching was mostly memorizing. For instance, Mary had to memorize an entire page of Samuel Johnson's famous dictionary, giving the spelling, pronunciation and definition of each word in order. Poor Mary! She had little memory for subjects which did not interest her.

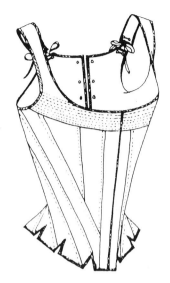

This is a fancy version of the kind of undergarment Mary wore at school. It was made of stiff material such as bone, and sometimes even steel and was tightly laced.

In addition, she had to wear a steel framework (called a busk) fitted around her chest under her gown during her lessons. Another steel piece attached to this fitted over her shoulders, drawing them back so far her shoulder blades touched in back. A third metal piece fitted in a semicircle under her chin. In this suit of armor she had to learn her lessons, as did all the younger girls in the school. This was supposed to keep them from slouching!

Mary never did adjust to life at the school, and at the year's end she returned home still writing and spelling poorly. Writing to her older brother who was at school at Edinburgh, she asked him to buy her something with the "bank-knot" she sent.

Later, when Mary was thirteen, she lived for a winter in the nearby city of Edinburgh. There she went to a day school to study writing and arithmetic. Finally she learned to write "a fair hand" and found that she liked arithmetic.

Home Life

Learning something new always interested her, but before she could follow the studies she grew to love Mary had some hard times at home. Whenever she gave her opinion on a subject she was "instantly silenced, although I often knew, and could have proved, that I was in the right." Mary was timid and hesitated to speak up. This shyness never left her though it was much less severe when she was older and famous.

Her relatives disapproved of her when she was a teenager because she read so much. Her father said, "We must put a stop to this, or we shall have Mary in a strait jacket one of these days."

It was commonly believed that girls and women could go mad if they exercised their minds too much. It was also believed, even by doctors, that too much study would affect the ability of women to bear children.

The pursuit of her studies was an area in which Mary was "determined and inflexible". Bit by bit she learned French, Latin and Greek. The subjects that fascinated her most were the ones that grew out of her love of nature. She had always had a keen eye for observing fish, birds, plants and the night sky.

Using a small globe that showed the constellations, Mary studied the skies from the north and south facing windows in her attic bedroom. Later, from reading a book on navigation, she learned that astronomy was not just stargazing, and that to make progress she would need to know higher mathematics to chart the constellations.

Puzzles and Clues

One day a friend of Mary's showed her a needlework pattern in a "ladies magazine." As the friend leafed through the pages, Mary's eyes were drawn to a number puzzle. Included with the numbers were the letters x and y.

Mary asked her friend, "What do these letters mean?"

Her friend replied, "Oh, it is a kind of arithmetic. They call it algebra, but I can tell you nothing about it."

Even though her friend had no information, Mary at least had the name of the mysterious subject. Later, when she was taking lessons in landscape painting (painting and needlework were considered acceptable subjects for girls to study), she heard her teacher talking to some of her classmates. He said, "You should study Euclid's Elements of Geometry, the foundation not only of perspective, but of astronomy and all mechanical science."

Another clue! Now she knew what books to ask for—books on algebra and geometry. About that time a tutor was hired for Mary's younger brother. This kind person agreed to buy the books Mary so desperately wanted and he did it on his very next trip to Edinburgh.

"Now," said Mary, "I had got what I had so long and earnestly desired. Before I began to read algebra I found it necessary to study arithmetic again, having forgotten much of it. I never was expert at addition, for, in summing up a long column of pounds, shillings, and pence in the family account book, it seldom came out twice the same way."

Mary read Euclid's work on geometry at night, after the rest of the family and servants went to bed. When the servants discovered that the candle supply was practically used up by Mary's nightly reading habit, Mary's mother told them to take away all her candles at bedtime.

Mary wrote, "I had, however, already gone through the first six books of Euclid, and now I was thrown on my memory which I exercised by beginning at the first book and demonstrating in my mind a certain number of problems every night, till I could nearly go through the whole."

Accomplishments

Studying mathematics was not the only thing Mary did. She spent hours every day at household chores and made and mended all her clothes, even her ball gowns.

About this time Mary and her mother went to Edinburgh to spend the winter. Here Mary made many social calls and went to dances, where everybody danced reels and other dances. Mary loved dancing and was never without partners. She was allowed to be accompanied only by gentlemen her mother knew. It was the custom for gentlemen to transport young women to the balls in sedan chairs. Mary often came home from a dance in bright daylight.

Mary had a small and delicate face and figure and wore becoming and fashionable gowns. She was very fresh looking and was called "The Rose of Jedburgh," Jedburgh being the name of the town where Mary stayed with relatives.

As always Mary loved to read, so much so that one day Mary's aunt, who lived with the family, said to her mother, "I wonder you let Mary waste her time in reading. She never sews more than if she were a man."

Mary disagreed. She wrote, "I was annoyed that my turn for reading was so much disapproved of, and thought it unjust that women should have been given a desire for knowledge if it were wrong to acquire it."

Mary loved music too and practiced the piano for hours each day. However, when her family obliged her to perform in front of company, her shyness returned and she played badly.

As a landscape painter she was accomplished and her work was passed around to friends of the family who encouraged her. Though Mary did not seriously consider a career as a painter she was, she wrote, "intensely ambitious to excel in something, for I felt in my own breast that women were capable of taking a higher place in creation than that assigned to them in my early days, which was very low."

Marriage

When Mary was twenty-four she married. Her husband was a distant cousin of hers named Samuel Greig and he was Consul to Russia. Before consenting to the marriage, Mary's father insisted that Mary would never have to live in Russia. There, stormy politics made life for foreigners dangerous because in 1804 Napoleon had declared himself Emperor and was leading his troops to battle with the Austrians and Russians.

Mary, who had grown up in a small Scottish village on the northern seacoast, moved with her husband to London where the Russian consulate was located.

Her husband did not forbid Mary to study, but he had no good words for her occupation. Like most other people, he thought that educating women and girls was foolish and useless.

During the next three years, two sons were born to the couple. Then, suddenly, Samuel Greig died. In those days people often died very suddenly. There were no drugs or innoculations against diseases such as cholera, pneumonia or influenza.

Widowhood

Mary and her children moved back to her parents' house. At twenty-seven she was a widow. She threw herself into her beloved studies.

For the first time in her life she did not have to meet with anyone's approval, for she was now independent. She resumed her old schedule of early morning and late evening study. She devoted the days to her children.

A Career

At this time the "ladies magazines" contained sewing and needlework patterns and recipes just as the *Ladies Home Journal* and *Woman's Day* do today. *The Ladies' Diary*, for instance, was a magazine that was published in England during Mary's lifetime.

Magazines and journals then had one major difference from those of today. Some of them contained problems and puzzles in mathematics. *The Ladies' Diary* was one of these. Mathematics was something new to people who were less educated, and women became as interested in the subject as men. It was not thought that mathematics was "over the heads" of women.

Mary began to work on problems given in one of the journals. She sent her solutions to the editor, Mr. Wallace. He was impressed with them and he sent back his own solutions.

"Mine were sometimes right and sometimes wrong," she wrote, "and it occasionally happened that we solved the same problem by different methods."

Finally Mary solved a problem in algebra and won a prize. It was a silver medal with her name on it.

This same Mr. Wallace became Professor of Mathematics at the University of Edinburgh. Mary told him she wanted to learn "the highest branches of mathematical and astronomical science" and he gave her a list of books which were written in French, Latin, and English. The list included books on algebra, physics, calculus, geometry, astronomy, logarithms and probability theory.

"I was thirty-three years of age when I bought this excellent little library," she said of herself. "I could hardly believe that I possessed such a treasure when I looked back on the day that I first saw the mysterious word `Algebra', and the long course of years in which I had persevered almost without hope. It taught me never to despair."

A Proposal

Mary received several offers of marriage. One offer, sent in writing as was the custom, listed the duties of a wife in such a narrow-minded way that Mary tossed it away. She knew what she wanted.

After a time a cousin named William Somerville proposed marriage and Mary gladly accepted. He and Mary were married for sixty years and they were very happy together.

William, whom Mary always called "Somerville" in her writings, was very proud of Mary's growing skill as a mathematician. He searched out books at libraries for her. Since he was very interested in the correct use of language, he re-copied her manuscripts before they were sent to the printer in order to correct any errors in spelling, punctuation and grammar. He was not jealous of her fame or her brilliance. He was, Mary wrote many years later, "one in ten thousand."

The First Book

Lord Henry Brougham, a publisher, realized that an English translation was needed of a new and important work on astronomy, written in French. He asked Mary to do the work. Because she had never studied at a university and had taught herself, she doubted her own ability.

She told her husband and Lord Brougham, "You must be aware that the work in question never can be popularized, since the student must at least know something of the differential and integral calculi, and as a preliminary step I should have to prove various problems in physical mechanics and astronomy. Besides, Laplace (the author) never gives diagrams or figures, because they are not necessary to persons versed in the Calculus, but they would be indispensable in a work such as you wish me to write. I am afraid I am incapable of such a task, but as you both wish it so much, I shall do my very best upon condition of secrecy, and that if I fail the manuscript shall be put into the fire."

She did not fail and her first book, *The Mechanism of the Heavens*, was published in 1831. In the work Mary published her own solutions to difficult problems set by Laplace. She gave clear accounts of experiments and gave examples. She wrote an introduction which was suitable for the ordinary reader.

The book was an instant success. It was used as a textbook for students at Cambridge University, the center for the study of mathematics in England.

Professor Peacock, a mathematician from Cambridge, wrote to Mary, "I consider it to be a work of the greatest value and importance, which will contribute greatly to the knowledge of physical astronomy."

After this she published three more books: *The Connection of the Physical Sciences*, *Physical Geography* and *Molecular* and *Microscopic Science*. Mary had discovered her niche—spreading knowledge about her beloved subjects. Her books were used by mathematicians, scientists and students. They sold well and were important in popularizing science.

This was a time when science was becoming a great influence upon the lives of ordinary people and many people were excited and curious about it.

Fame

Mary received many honors, honorary degrees, and medals. A statue of her was placed in the Hall of the Royal Society in London. In 1834, The Royal Astronomical Society named her and Caroline Herschel as its first honorary female members. A new sailing ship was called the *Mary Somerville* and a copy of her statue was placed on the ship's prow as a figurehead.

Because Mary was a woman working in a field where almost no women were to be found, she was especially well known. Since she successfully combined marriage, family, and a scientific career, she helped the cause of other women interested in mathematics and science. She accomplished this at a time when women were not even admitted to universities.

As she grew older she continued to be very energetic. As long as she lived she rose early, studied until early afternoon, then put her books away. From then on she took care of her household and went out to visit friends. Of course she, as everyone of her social class, had servants to make this schedule possible.

Mary Somerville as a mature woman.

Family Life

Mary and William became interested in minerals. They invested their money in gems such as rubies, sapphires, topaz and amethysts. In the evenings they liked to take their collection out of the cabinet and arrange the gems or just admire them.

Mary had three daughters. The oldest, a child of unusual talent, died and Mary grieved for a long time. The two surviving daughters, Mary and Martha, never married but stayed in their parents' household.

In the 1840's when Mary was in her sixties the family moved to Italy. William was ill and the doctors advised a warmer climate. Woronzow Greig, Mary's son by her first marriage, lived in England with his family. Mary and Woronzow were very close. They visited often and exchanged letters. Woronzow handled Mary's legal and publishing affairs.

As her career progressed, Mary turned from mathematics to science. She later regretted this decision since she believed her greatest gifts were in mathematics.

A Pet

All her life Mary loved birds, especially songbirds. A mountain sparrow was her pet for eight years and when Mary sat in bed in the morning, writing and reading, the bird would fly in the room and perch on her arm.

She had a horror of killing animals, either by hunting or by vivisection, even for scientific experiment. She petitioned the lawmakers of England and Italy to outlaw these practices, which she believed were cruel. When Mary was a very old woman she wrote that she believed that God, in his great mercy, would provide an everlasting home for helpless animals, as well as for human souls.

Old Age

Mary Somerville lived to be 92 years old. By that time few of her friends were still alive.

"I am nearly left alone," she wrote. Indeed, her beloved husband and son had died before her.

Her daughter Martha published her memoirs a couple of years after her death. Martha wrote that Mary worked up to the very day she died. She was quite deaf, but her eyes were still good. They were so good she could pick out the threads of her needlework canvas without using glasses.

Mary never lost the power of her mind nor her memory. Even in her nineties, she could work difficult problems in algebra with the same joy and determination she knew as a young girl.

Shade all multiples of 9 to enhance this picture of Mary Somerville.

Drawing by Analee Nunan.

Activities _____

?

● **What is a digital sum?** But first ... are you sure *what a digit is?*

A All numbers, no matter how huge, are made up of combinations of only ten digits. These digits are 0, 1, 2, 3, 4, 5, 6, 7, 8, 9.

examples— 351 has three digits ... 3, 5 and 1; 603801 has six digits ... 6,0,3,8,0,1.

note— The *value of the number is set by the placement* of these ten digits within the number.

?

● **What is a digital sum?**

A The *digital sum* is the sum of the digits in a number. If the initial sum is more than one digit, add the numbers in the resulting sum again until the sum of the digits is a single digit. That single digit is called the digital sum of the original number.

note— *If the digital sum, is 9, then the original number is a multiple of 9.*

examples— Both 351 and 603801 are multiples of 9. That can be checked by dividing each of these by 9 and seeing that there is no remainder. The easier digital sum method shows...

The digital sum of 351 is 3+5+1 which equals 9.
The digital sum of 603801 is 6+0+3+8+0+1=9+9=18 then 1+8=9, and we say the digital sum of 603801 is 9.

? ● What is a common multiple of three and five?

A Numbers that are *common multiples* of 3 and 5 are numbers that are multiples of 3, and at the same time are multiples of 5.

Short Cut—

• The last digit of all numbers that are multiples of 5 is either five or zero. Any number that ends in anything but five or zero fails that test.

• All numbers that are multiples of 3 leave no remainder when divided by 3, and have a digital sum that is 3 or 6 or 9. The digital sum test is easy to do.

• All numbers that pass both tests are common multiples of 3 and 5.

Solutions are given at the end of book in the section SOLUTIONS TO ACTIVITIES.

MARY'S SYSTEM—

Mary Somerville used an approach to her work that is useful today. If she couldn't find the key to unlock a difficult problem she stopped working and turned to the piano, her needlework, or a walk outdoors.

Afterward, she returned to the problem with her mind refreshed and could find the solution.

If she could not understand a passage in her reading, she would read on for several pages. Then, going back, she could often understand what was meant in the part which had been confusing.

When she was trying to master geometry she worked out problems in her mind at night in bed, beginning with the simplest problems and proceeding to more complex ones.

Her success as a mathematician may have depended in part upon these basic habits.

'ON THE WAY TO THE BALL' — Shade all common multiples of 3 and 5.

Drawing by Analee Nunan.

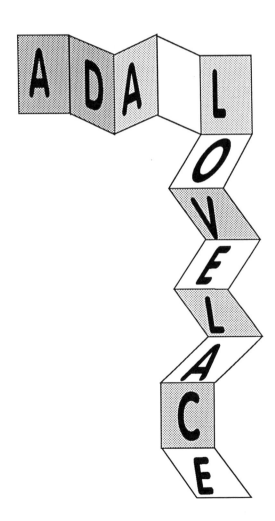

ADA LOVELACE
1815-1852

"Pussy cat, pussy cat, where have you been?

I've been to London to look at the queen.

Pussy cat, pussy cat, what did you there?

I frightened a little mouse under her chair."

Whether she was timid as a little mouse or courageous as a lion, only Ada, in her 17-year-old heart knew, for she had learned to keep the feelings in her heart secret. Her mother called her "the young Lioness" when she was presented to the Queen in the London season of 1833.

Ada was on the brink of entering the adult world, for she was meeting King William IV and Queen Adelaide at the Court of St. James. Her full name was Augusta Ada Byron, and she was thrilled with the splendor of the company and of the Palace itself. She had attended few parties in the past, but in the winter of 1833 she went to many balls and dinners. This was her "coming out" season.

At these gatherings, for the first time in her life, she met many important and famous people. Many of them were not only famous, but they were intelligent, accomplished and witty. Ada yearned to be part of their group.

Ada at seventeen about to go out into the world. Sketch by Analee Nunan.

Lord Byron, her father

In a way you could say that Ada had been born into such a group. Her father, George Gordon, Lord Byron, was a poet who was known throughout the world.

Lord Byron had come to London when he was a young man. He became known in society quickly because of his poetry. A handsome man, he was also talked about in social gatherings because of his affairs with many rich and beautiful women.

Lord Byron as a young man.
Courtesy of John Murray (Publishers) London.

Besides being a playboy, Byron wrote biting criticism of the weaknesses of English society. He attacked corrupt politicians and their lies. He stood up for factory workers when they rioted because of terrible working conditions.

Byron flirted, threw away his money, traveled all over the Continent and had many adventures. Finally, one day, home from his travels and attending still another party, he met an attractive, very proper young English woman named Annabella Millbanke.

Annabella had been brought up in a comfortable upper class English country home. She believed in such values as duty, obedience and good works. She was a gifted student of mathematics and literature and had read Byron's poems. Even more important, Anabella Millbanke had a generous dowry which would enhance the financial position of any man she married. Although Byron was part of the aristocracy he was in a terrible position financially. Because of her dowry, Anabella was considered an excellent marriage prospect. Because of his level in society, as well as his personal attractiveness and fame, Byron was also considered a "good catch".

Unlikely as it seems, these two people fell in love. They married, but they did not "settle down." They moved about from place to place. Lord Byron treated Annabella very cruelly at times and made it clear that he was sorry he had married her. She, in turn, prayed for his soul, and pointed out his problems in long-winded speeches.

One year was all the marriage lasted. A week after the birth of their daughter, Ada, husband and wife separated forever.

Lady Annabella went to her parents' home with Ada. Lord Byron stormed away to Italy.

Lady Byron at the time Ada was a young girl.
Courtesy of John Murray (Publishers) London.

Ada! sole daughter of my house and heart?
When last I saw thy young blue eyes they smiled
. . . when we parted,—not as now we part
But with a hope . . .

Byron wrote these lines about Ada when she was one year old. He was never to see her again, for Byron died in Greece, supporting a war for Greek freedom from the Turks.

"The Separation," as everyone in English society called it, was gossiped about for many years. Lady Byron kept most of the gossip from reaching Ada's ears as Ada grew up. But though everything was "hush-hush," the young girl felt something. When the news of her father's death reached the family, Ada was eight years old.

Ada was caught between the two extreme types of her parents' personalities. This conflict took its toll. Although as a child she had been described as cheerful and robust, she suffered a kind of nervous attack when she was eight years old. She began to suffer from severe headaches that affected her eyesight and made reading difficult for several months. At thirteen her legs became paralyzed, and it was several years before she could move about normally.

This picture of four year old Ada was in a tiny locket sent to her father by her aunt.
Courtesy of John Murray (Publishers) London.

Ambition

Ada overcame these illnesses and actually went on to became active in gymnastics, dancing and her great love—horseback riding. Her letters show her to be a high-spirited person despite being constantly surrounded by her mother's advisors, who preached at the teenage girl about almost everything.

Ada was gifted in many ways. She loved music and was a promising and versatile musician. She played the harp, violin and piano. She enjoyed mathematics and loved mechanical things. When she was eight and nine years old, she built and played with toy ships and boats.

When Ada was seventeen and studying algebra and astronomy, which she had mostly taught herself, she wrote playfully to a friend, "So this you see is commencement of `A Sentimental Mathematical Correspondence between two Young Ladies of Rank' to be hereinafter published no doubt for the edification of womankind. Ever Yours Mathematically."

During her "coming out" season in London, Ada was looking for others to share her great love for mathematics, music, and riding, and anything else that was interesting and new. Most of all, she wished to meet the famous Mrs. Somerville. Mary Somerville was living in London at this time. She had just published *The Mechanism of the Heavens* on mathematical astronomy. This book was being read with fascination by the educated people of the day. Ada dreamed of becoming a famous mathematician like Mrs. Somerville.

Babbage and the Difference Engine

One of the brightest stars of the London scene was inventor Charles Babbage, who had studied at Cambridge University. Babbage traveled widely and involved himself in many scientific projects. He loved to see and be seen at the best parties and balls. At one of these, Ada was introduced to him. She was delighted. He invited Ada and her mother to see his pet project, an early computer.

Charles Babbage (1792-1871) as an old man, shortly before he died. From the *Illustrated London News, November 4, 1871.*

His computer project, in fact, obsessed him. He called it the Difference Engine and it had many moving parts. This machine would be able to produce, with awesome speed and accuracy, tables for astronomers and navigators to use. Up to this point, these tables could only be computed tediously by people who multiplied and divided numbers by hand. The accuracy of these calculations, of course, was limited. The

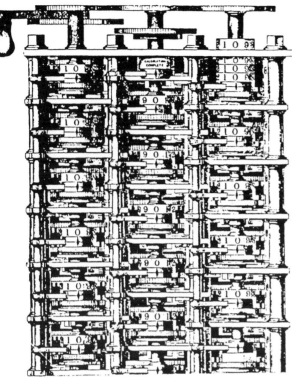

The Difference Engine, one of Babbage's famous calculating machines. This machine is often on exhibit in the Science Museum in London, England.

Difference Engine added numbers instead of multiplying them, and the most powerful thing about the machine was that it could add many numbers very quickly and accurately. The tables which were computed by the Difference Engine were badly needed, because shipping was England's leading industry. So Babbage received two grants of money from the government to build his machine.

Although Babbage was a genius, he was not a very good manager. When the money ran out the workmen, whose skills he needed to produce all the Engine's precision parts, quit for lack of wages. The project was never completed.

When Ada saw the Difference Engine she was enthralled. The evening after Babbage's demonstration to her and her mother, Ada went to the Queen's Ball and again conversed with Babbage. Thus began a professional tie and a friendship which lasted throughout her life.

Ada wanted to work with Babbage and began studying differential equations so that, as she wrote to him, "at some future time . . . my head may be made by you subservient to some of your purposes and plans." But it was to be some time before she would begin working with Babbage.

Meanwhile, Ada wrote about her work to another well-known mathematician, Augustus DeMorgan. DeMorgan was a kind of tutor to her.

He wrote to Ada's mother, Lady Byron:

I feel bound to tell you that (Ada's) power of thinking on these matters has been something utterly out of the common way for any beginner, man or woman. Had any young beginner, about

Augustus De Morgan, the eminent mathematician who recognized and encouraged Ada's mathematical talents.
From *Memonirs of Augustus De Morgan* by Sophia De Morgan, 1882.

to go to Cambridge, shown the same power, I should have prophesied (that he would become) an original mathematical investigator, perhaps of first-rate eminence.

Ada Lovelace at nineteen soon after her marriage to Lord King, later to become Earl of Lovelace. Sketch by Analee Nunan.

Romance

Finally, Ada was able to meet her heroine, Mrs. Somerville. The older woman became very fond of Ada. Ada was invited to attend concerts and plays and even to stay overnight with the Somervilles. Woronzow Greig, Mrs. Somerville's son by her first marriage, and her two daughters, Martha and Mary, made up their group.

One day a Cambridge friend of Woronzow's, whose name was Lord William King, came to call.

At this time Ada was a lively young woman who looked like her handsome father. She had "large, expressive eyes and dark curling hair."

Lord King and Ada fell in love. They became engaged to be married. Eleven years older than Ada, and of a serious nature, he wrote love letters that were full of tender feeling. Long engagements were not in style and the two were quickly married. Ada was nineteen years old and her husband was thirty.

Home & Family

Their first child, Byron, was born a year later. The following year a daughter, Annabelle, was born, and soon after, a third child, Ralph. From then on, Ada's health was never good for long. The symptoms of mystifying illness that she had shown as a young girl—symptoms that were beyond the skill and understanding of the doctors of her

Ashley Combe, Somerset, the second home where Ada and her husband lived. Sketch by Analee Nunan.

day— showed up again. She had digestive trouble and developed kidney disease as well as heart trouble and asthma. Between these attacks, however, Ada felt optimistic and continued to search for ways to pursue her mathematics career.

As a home manager, Ada was not successful. She loved her children, but hated the interruptions they caused in her work. Although she and her husband were aristocrats, and much of the time kept servants, tutors, and governesses, they were not wealthy, and sometimes day-to-day duties fell to Ada. So eager was she to rid herself of these domestic chores and make time for

Ada Lovelace's three children, at different ages. The eldest, Viscount Ockham (Bryon) is thirteen and is shown wearing a midshipman's uniform. The younger son, Ralph, is 10 years old in this picture. He later became the second Earl of Lovelace.. Courtesy of John Murray (Publishers) London.

her mathematical work that she begged her mother, who was always ready to take over, to take the children off her hands.

Ada's daughter Anne is probably about thirty years old in this sketch, married and known as Lady Anne Blunt. Sketch by Analee Nunan.

Three years after their marriage, Lord King inherited another title. He became Earl of Lovelace and Ada became the Countess of Lovelace. Although this title brought them no more money, Lord Lovelace was given a place in the House of Lords in Parliament.

The Analytical Engine

Meanwhile, Babbage was working on plans for a far more complex machine which he called the Analytical Engine. Indeed, the large-scale electronic digital computers of our own century copy the logical structure of Babbage's Analytical Engine, which he thought up in the 1830's.

Ada's enthusiasm for this computer matched Babbage's. Another admirer of Babbage's, L. F. Menabrea, wrote an article in French describing the Analytical Engine and its principles of operation. Since Babbage had not written about his machine, this work filled a real need. Ada translated Menabrea's article into English and, in the process, expanded the contents in important ways. Her notes, three times the length of the original article, set down in concrete terms the powers as well as the limitations of the machine.

A Published Work

The work was published, and Babbage proudly distributed copies to the leading scientists of the time. He was tremendously impressed with Ada's paper, and asked her why she had not written a separate original article on the subject. She replied that the thought had not occurred to her. In fact, even signing her own work caused her great anxiety.

Although her paper was clearly the work of an expert, it was also the work of a woman and a woman of rank at that. For such a woman to publish a scientific article was highly unusual. It was even more unusual when the area in question was such an "unfeminine" one as mathematical computation. After much indecision, Ada signed her paper "A.A.L.", using her initials only. It was many years before the author's identity was commonly known.

Babbage proposed using punched cards to put data into the Analytical Engine. This was similar to the clever method invented by J. M. Jacquard, who used punched cards to control the sequence of threads in a loom in order to weave fabulous fabric designs.

Ada saw the same kind of beauty in the Analytical Engine, and she wrote: "We may see most aptly that the Analytical Engine weaves algebraical patterns just as the Jacquard loom weaves flowers and leaves."

Ada's paper provided the public with the best account of the Analytical Engine, an account which Babbage saw was far clearer than he himself could possibly have written. As it turned out, this paper was the highpoint of Ada's career. No one knew exactly why she never went on. Perhaps it was because her health was bad, and she had difficulty focusing her attention on intellectual problems.

Even as she was seriously working with Babbage, Ada participated in the social life that was typical of people in her position and class. Here she is depicted at twenty-seven, dressed up in a fancy costume. Drawing by Analee Nunan.

Music Ambitions

Ada predicted computer music a whole century before it was actually produced!

She wrote to Woronzow Greig: "I am not dropping the thread of science, Mathematics, etc. These may still be my ultimate vocation, although it is likely to have a formidable rival — musical composition."

Ada, the promising musician, in her paper on the Analytical Engine, suggested that the computer might be used to compose music. "If," she wrote, "the fundamental relations of pitched sounds in the science of harmony and musical composition were susceptible of sufficiently precise formulation."

Health & Money Troubles

But she was really not well enough to work at either music or mathematics. She wrote to Woronzow:

"There is in my nervous system such a want of all ballast and steadiness. I am just the person to drop off some fine day when nobody expects it. Do not fancy me ill. I am apparently very well at present. But there are seeds of destruction within me. This I know."

Ada swung between feelings of doom and exuberant joy and optimism. Her husband was different. His life was wrapped up in the management of his estates. Developing and maintaining them was an expensive matter, and Lord Lovelace was not generous with money for other things.

He gave Ada 300 pounds yearly. This amount was part of her marriage settlement, a substantial 16,000 pounds. The entire sum was placed in a trust for her and would come into her hands only upon the death of her mother, Lady Byron. A much larger fortune, the bulk of her mother's large estate, would go directly to her husband, Lord Lovelace, according to the inheritance laws of the time.

As the only child of a very rich woman, Ada resented the small amount she was expected to live on. Her husband, she pointed out, would receive 7000 pounds per year— a sum, she wrote, "he is to enjoy to my exclusion!" Ada, it turned out, had reason to be concerned about money.

Playing the Horses

Ada loved horses. When she was seventeen, she tried to avoid going on a vacation to Brighton resort because she had just gotten a new horse named "Sylph," and she wanted to ride instead. Lord King too was an excellent rider, and the two of them loved to ride together.

Ada's passion for horses was now combined with another passion—gambling. Ada was a compulsive gambler and could not quit once she started. She began to bet larger and larger amounts of money on horse races. She lost heavily. Part of the problem was her pride in her mathematical genius. She kept working on new formulas to help her figure the odds and pick a winner.

Since Ada had no control over the family money and her husband did not approve of her large-scale gambling, Ada turned to "shady" money lenders to help meet her debts. These people blackmailed her, probably with the threat of telling her mother. Desperate for money, Ada pawned the family diamonds, not once but twice. Yet the debts kept piling up.

Tragedy's Path

Bad luck seemed to attract more bad luck, and Ada's illnesses returned with renewed force. She began to bleed internally. A tumor was found in her uterus. Cancer!

Ada was crushed though she did not give up hope. Between bouts of intense pain, her spirit showed itself again and again. She had her bed moved near her beloved piano, and each day she played. Sometimes she played duets with Annabelle, her daughter, who was now fifteen years old.

Ada's mother moved into the Lovelace home, and from that time onward, visitors were not permitted to see Ada. Ada could not leave her bed. Babbage in particular was kept away. Ada's mother felt bitter that it was his maid who had carried Ada's bets to the bookmakers.

Ada, at the piano, in 1852, painted when she knew she was dying. The sessions were often interrupted by severe pain, but Ada insisted on being painted by the son of the artist who had painted her father. Courtesy of John Murray (Publishers) London.

Ada herself never lost her affection for Babbage. Lord Lovelace wrote in his journal eleven months before her death: "Babbage was a constant intellectual companion and she ever found in him a match for her powerful understanding, their constant philosophical discussions begetting only an increased esteem and mutual liking."

A rendition of a sketch Ada's mother drew when Ada was dying.
Courtesy of John Murray (Publishers) London.

The Final Resolution _____

Now, of course, Lady Byron knew all. Ada's gambling losses could no longer be kept secret. Lady Byron paid out 5000 pounds to cover her daughter's losses.

Daily, Lord Lovelace struggled with his grief and with Ada's creditors.

Ada's agony stretched out for nearly four months while her helpless family waited. Finally, two weeks before her 37th birthday, Ada died.

"By her own wish," a descendant wrote, "they carried her to the old Newstead country, and laid her by the father whom she had never known."

Ada's Influence _____

In the brief time she lived, Ada Lovelace distinguished herself as a mathematician, and has even been called the inventor of computer programming. She used her gifts against tremendous obstacles and showed a spirit that would not be crushed.

As to Babbage and the influence of his Analytical Engine, his biographers wrote, "The Analytical Engine was never built, though Charles Babbage lived nearly another two decades. . . The Menabrea/Lovelace paper remains as the sole witness of the power and scope of the ideas of Babbage's Analytical Engine. These ideas lay dormant for another century."

In a similar way, the Lovelace paper remains the sole witness of the power and scope of Ada Lovelace's special genius. Now, a century later, we marvel at her early contribution to today's programming of computers.

Activities

> • Complete the triangular numbers listed below.

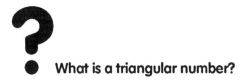

● What is a triangular number?

A   The **triangular numbers** are the sums of the counting numbers 1, 2, 3, 4, . . . (See page 104 **Pascal Triangle**, for more about this special set of numbers.) To calculate the numbers, use the information that T2 = N + T1, where T2 is the next triangular number; N is the first (1), second (2), third (3), etc. counting number; T1 is the previous triangular number.

example — $T_{(n=5)} = N + T_{(n=4)} = 5+10 = 15$

$T_{(n=6)} = N + T_{(n=5)} = 6+15 = 21$ etc.

Use this list of triangular numbers (T_2) to complete the portrait of Ada Lovelace on the next page.

N	+	T_1	=	T_2
1		0		1
2		1		3
3		3		6
4		6		10
5		10		15
6		15		—
7		—		—
8		—		—
9		—		—
10		—		—
12		—		—
13		—		—
14		—		—
15		—		—

Shade all triangular numbers to enhance this picture of Ada Lovelace.

Drawing by Analee Nunan.

Programming Without A Computer _____

Make two copies of this maze.

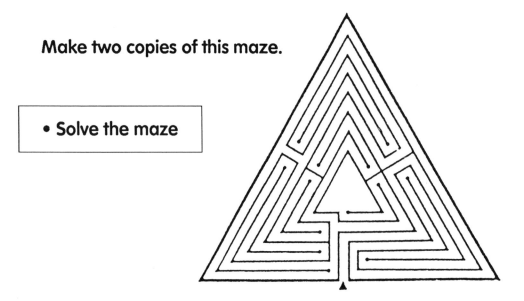

• Solve the maze

• Program the solution

Use only the following commands ...

G   Go forward until you bump your head

R90   Turn right 90 degrees

R120   Turn right 60 degrees

L90   Turn left 90 degrees

L120   Turn left 90 degrees

• Test it on a friend

Give your friend a copy of the maze.

Give your friend a copy of your program.

Can your friend solve the maze by following your program?

note— A program is a series of steps that tells you exactly what to do. For example, in a figure that looks like this, **s** and starting at **s**, the program **G, R90, L90** would take you along the path **s** . The program **G, L90, R90** would look like **s**

Idea adapted from Tim Erickson's *Off & Running.* Copyright © 1986. EQUALS, Lawrence Hall of Scince, Berkeley, CA.

For Penrose's Amazing Maze you only need three commands:

  G,

  R90, and

  L90.

Again, SOLVE! PROGRAM! TRY ON A FRIEND!

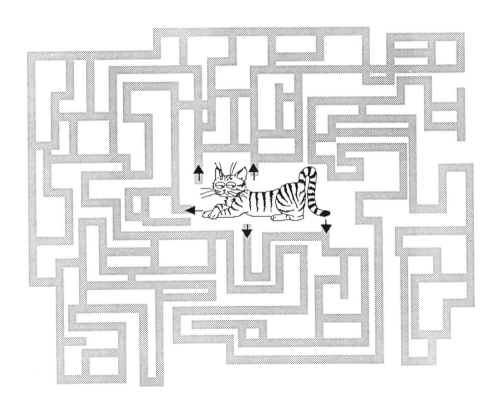

Penrose's Amazing Maze was reproduced from **The Children's Mathematics Calendar 1992** by Theoni Pappas. Copyright © 1991. Reprinted by permission of Wide World Publishing/Tetra, San Carlos, California.

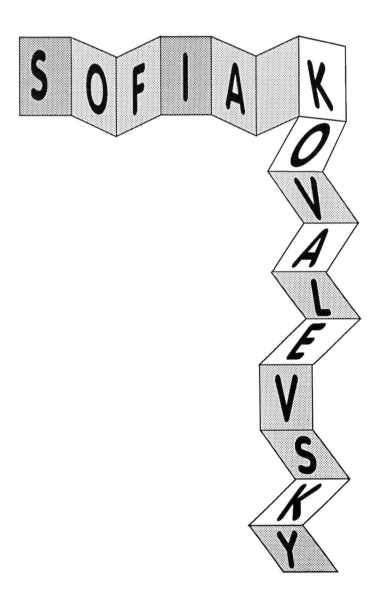

SOFIA KOVALEVSKY*
1850-1891

**"It is impossible to be a mathematican
without having the soul of a poet." —Weierstrass**

Sofia was eight years old when her family moved from Moscow, where she was born, to Palibino, a large estate in Vitebsk. In winter, wolves howled around the estate, and deep snow buried the road to St. Petersburg, the nearest large city. Pine forests protected the large house from the winds that blew over miles of plain.

Indoors, the family lived in a private world. The head of the household was General Krukovsky, Sofia's father. Her mother, who was much younger than the General, liked to read novels, paint and play the pi-

This house in Palibino where Sofia spent much of her unhappy childhood. A museum dedicated to Sofia Kovalevsky's life and works is being created at Palibino. When complete, the house and grounds will look just as they did when she lived there. The target date for its completion is 2000—the 150 anniversary of Kovalevsky's birth.

ano. Aniuta, Sofia's sister, was six years older, and Fedya, her brother, was three years younger. Sofia's maternal grandfather had been an eminent mathematician and astronomer.

A Strict Schooling

From the time Sofia was five years old until she was twelve, she had only one teacher, an English governess. This woman did not understand Sofia's passionate nature. She tried to press Sofia into a mold labeled "proper English miss"—a mold that would produce a young woman who played the piano, had perfect manners in company, and was neat, tidy and prompt. She would not let Sofia read the romantic novels and poetry she loved, and she urged long winter walks on her, to improve circulation! Sofia escaped only when the temperature fell below 10 degrees.

---

*Over the years her name has been written in several different ways. For your reference some of these are: Sofia Kovalevsky, Sofya Kovalevsky, Sonya Kovalevskaia, Sonia Kovalesvskaya.

Sofia's mother, Elizabeth Federovna Kowin-Krukovsky. Mother and daughter were too different to ever be close.

The governess held total power over young Sofia's life. Her father felt it was not his place to make decisions about Sofia's education, and her mother was not strong enough to go against the governess' rules. Sofia felt her parents did not really care about her.

The Krukovsky house was quite large, and each section was occupied by a part of the family. Sofia's father lived in the tower, her mother and sister lived upstairs in the main apartments, Fedya and his tutor lived in a separate upstairs wing, and Sofia and her governess lived on the ground floor.

Sofia's father, General Kowin-Krukovsky. Sofia was said to be her father's favorite.

More and more, the person to whom she showed her affectionate heart was her teenage sister, Aniuta. Because Aniuta and the governess despised one another, Sofia had to sneak when she wanted to see her sister.

Studying the Wallpaper

All of the rooms were decorated with fancy imported wallpaper. Only one was left unfinished. In this room the imported wallpaper had run out, and old papers were used to cover the walls. The papers were lecture notes on differential and integral calculus which the General had used in his student days. Sofia was alone a great deal, and she amused herself by staring at the squiggles on the papers. Years later, she studied calculus in St. Petersburg. She caught on so quickly her teacher remarked that she must have known the subject all her life. In a way she had.

Sofia was a brilliant student. After her governess left, she taught herself physics. But in order to do this, she found she would need to learn trigonometry. So she taught herself!

One friend of the family was a physicist. On a visit to the Krokovskys, he questioned Sofia about her studies. When he discovered what she was actually learning on her own, he went to the General and convinced him that Sofia must be allowed to study mathematics in St. Petersburg. The General finally consented.

The World Outside

Sofia was growing up at a time of great social and political upheaval in Russia. In 1861, Czar Alexander II gave the serfs their freedom. Ideas about equality for all spread even into the country and to the estate at Palibino.

A young man from the nearest village struck up a friendship with Aniuta. He had been living in St. Petersburg and had learned the new political ideas. He brought magazines and journals that told of the struggle all over Europe for freedom from hunger, poverty and harsh rule. They told of the serfs' desire for equality, of women's desire for higher education, and the people's desire for political power.

Many sons and daughters of the aristocracy, like Sofia and Aniuta, criticized the old-fashioned ideas of their parents. They were just beginning to understand that they did not wish to continue the traditions of marrying early and settling into the role of upper class landowners.

Sofia, at fifteen, spent the winter in St. Petersburg with her mother and sister. There she got to know the famous Russian writer Dostoevskii, who was a regular visitor courting her sister. Gradually Sofia developed a crush on him that caused her many painful hours.

Anuita, the sister whom Sofia so highly regarded.

A Wild Plan

Aniuta was now a grown-up young woman, longing to be free of her father's strict rules. Yet she was stuck at Palibino. She asked her father if she could study at the University of St. Petersburg and her father said no.

At this point, around 1860, the University had just opened classes to women, although women could not earn degrees. A couple of years later, when students protested in order to gain more liberties, the University closed. When it reopened, women were no longer permitted to attend classes there, nor anywhere else in Russia.

Sofia, like Aniuta, wanted to leave home. She wanted to study science at a university. Wherever Aniuta went, Sofia decided, she would follow.

Together with Sofia and a friend, Aniuta devised a wild plan. The young women would ask a young man who shared their political beliefs to marry one of them! Such a marriage would set them free. Once they were married, the couple, plus their friends, would set off to study at a university in Germany or Switzerland. In these countries women had a better chance of being admitted to universities.

Such a marriage, it was understood, would be in name only. After their arrival in the foreign city the husband and wife would live apart.

The first young man they tried to interest in their plan turned them down. But "no" did not discourage these three. Next, they tried a promising student of geology. His name was Vladimir Kovalevsky, and he said "yes."

Sofia, at the age of eighteen.

There was one problem: he wished to marry Sofia, the youngest. He wrote to his brother, "Despite her eighteen years the Sparrow (Sofia's nickname) is extremely well-educated. She speaks foreign languages as fluently as her own. She studies mainly mathematics, is now tackling spheric(al) trigonometry and integrals. She is as busy as a bee from morning till night and still is lively, sweet, and very pretty."

To this bold marriage proposal the General said, "No"! It was unheard of that a younger sister marry before the older. But Sofia had seen her chance to escape and follow her dream of getting a higher education.

Sheer desperation gave her the courage to do what she did next. She, who was never allowed outside her home unchaperoned, slipped away to join Vladimir in his apartment in the nearby village. According to the rules of the society in which they lived, such behavior was equal to eloping. She left a note for her father, telling him where she was going and why. He followed her there. As she had hoped, her father now allowed their engagement.

Sofia and Vladimir were married with her family's blessing, and within six months, the couple was living in Germany.

Higher Education

During the two years spent at Heidelberg, Germany, Sofia's closest friend was Julia Lermontov, a chemistry student. Lermontov later wrote about Sofia during these years: "She was just eighteen, but looked much younger. Small, slender, with a round face and short curly chestnut hair, she had very mobile features. Her eyes, especially, were exceedingly expressive—sometimes bright and dancing, sometimes dreamy and full of melancholy, ... a mixture of childish innocence and deep thought... She took no pains about her personal appearance or dress... a trait which remained with her to the last."

Sofia and her friend Julia, were the first women at Heidelberg University. Russian women of their generation were the first to open higher education to women in continental Europe.

According to plan, Sofia became a student in mathematics at the university, and began her climb to fame and honor and, along the way, to tragedy and heartbreak.

Sofia's was a complex personality. On the one hand she was a daring idealist who was concerned with improving human life. She also loved society and its honors. On the other hand, she was a lonely scholar living for her work, avoiding other people.

Sofia always felt a hunger for love. Her own life, she believed, was terribly lacking when she observed the lives of her friends. As in her childhood, she now felt she was somehow outside circles of love.

Sofia moved to Berlin in order to work under the famous mathematician, Karl Weierstrasse. Here she spent four years, finally receiving her degree in mathematics from the University of Gottingen in 1874.

During this period she worked in almost complete solitude, often sitting for hours in her room trying to solve problems. When she succeeded, she would rise from her desk and pace the floor talking to herself, walking faster and faster, laughing and finally breaking into a run!

She certainly did not possess the disposition of a Mary Somerville. Rather, Sofia was nervous and withdrawn. She saw no one for days. She went neither to parties nor dinners.

Vladimir, her husband, also lived in Berlin during this period. They were still not living together. When he visited Sofia, they would go for long walks. This was her only recreation.

Sofia frequently returned to Russia. From there she once wrote, "I feel released from the prison in which my best thoughts were in bondage. You cannot think what suffering it is to have to speak always foreign languages to your friends. You might as well wear a mask on your face."

Vladimir Kovalevski, Sofia's husband in name only, during their Berlin years. Because they lived apart and because of their studies and limited funds, they rarely saw each other in Berlin. This was a lonely time for Sofia.

Sofia with Fufa , born in 1878. In 1882, Sofia sent Fufa to Russia after a serious illness. The child lived there first in the care of her uncle, then with her mother's friend, Julia Lermontova.

Married Life

After five years of a marriage that was a mere business arrangement, Sofia and Vladimir became lovers. True to her passionate temperament, Sofia threw herself into married life, hoping the marriage would succeed as a love match. She also hoped it would produce the happy family life she longed for.

A daughter was born who was also named Sofia. But she was known by her nickname, Fufa. Sofia put aside all of her studies to care for little Fufa and to promote her husband's scientific work. She yearned to be a typical wife and mother. But she was not typical. When Vladimir, then a lecturer in geology at the University of Moscow, began to lose interest in science, Sofia wrote his lectures for him. Instead of attending to his job, Vladimir plunged into business schemes. They failed, one after another.

Sofia believed she had the power to see the future. One night, she had a nightmare that a grinning monster was stamping Vladimir beneath its feet. She was terribly frightened. Later, she realized that the dream monster was Vladimir's business partner. This man cheated Vladimir out of a large sum of money. Through him all the family money, including Sofia's small inheritance from her father, was invested in Russian oil refineries. The refining company was later accused of fraud, and Vladimir was threatened with a lawsuit. While this was happening, the partner convinced Sofia that Vladimir was in love with another woman.

Sofia felt rejected. She left Russia with her daughter and returned to Berlin. There, in a hopeless state of mind, she threw herself once again into mathematics.

Vladimir committed suicide. Sofia was just thirty-two years old, and a widow. The death of her husband was a great blow. She never stopped feeling somehow responsible for his sad end.

Professor Sofia

In the next ten years, Sofia gained recognition and respect as a mathematician. Her teacher, Weierstrasse, introduced her to Gösta Mittag-Leffler, the great Swedish mathematician, who took up her cause. Mittag-Leffler got her a job as mathematics professor at the University of Stockholm in Sweden. This was a tremendous step forward for Sofia, and for women of science generally!

Sofia in 1885—a professor squared!! As a widow, in nineteenth century Europe, Sofia had much more independence than a single or married woman.

The first year, many parties and balls were given in her honor. Swedish girl babies were named "Sofia" after her. After all, she was the only woman professor in the country!

Gösta Mittag-Leffler (1846-1927) brought mainstream European mathematics to Scandinavia. Because of his support, Sofia Kovalevsky became the first woman in modern times to hold an appointment in a European university.

Sofia with Anna Carlotta Leffler, Swedish friend, author, sister of Mittag-Leffler, and well-known advocate of women's rights.

But Sofia had enemies as well as admirers. A famous Swedish writer, August Strindberg, wrote a scornful article about her appointment.

Sofia repeated what he wrote in a letter to a friend. "She proves, as decidedly as that two and two make four, what a monstrosity a woman professor of mathematics is, and how unnecessary, injurious and out of place she is."

Nevertheless her career went so well that she was given a five-year contract in 1884 and asked to become a professor of mechanics, a branch of physics. When she accepted this second job, she joked, "Now I have become a professor squared!"

Mittag-Leffler's sister, Anna Carlotta, a well-known writer, and Sofia became good friends. Sofia now began a second career as a writer. Her themes were mainly about her childhood in Russia.

Both women were feminists and held high hopes for the future of women in the world. The "woman issue," as it was called in Sweden, was a subject of hot debate at that time.

The two women began to work on a drama in two parts, called *The Struggle for Happiness: How It Was, How It Might Have Been*. Sofia was so excited about this work that she gave up mathematics for a time. She and Anna Carlotta dreamed about traveling to Germany and France to meet literary and theatrical stars and prepare for their coming fame.

They promised to write each other's biographies. With that in mind, they both began saving letters and documents. Anna Carlotta kept her promise. After Sofia's death, she wrote a biography of her dear friend's life.

It wasn't long, however, before Stockholm began to seem like a boring country town to Sofia, and one that was desperately cold in winter. Sofia longed to live in a dazzling world capital like Paris. But she needed a job, and women mathematicians were not wanted in most universities. In fact, rarely were women allowed to be students, much less paid teachers. Sweden was the only country in Europe, except for Italy, where there were women professors in any subject.

Sofia tried again to find work in Russia. After years of being ignored in her native country, she was made an Associate of the Russian National Academy of Sciences. So, with hopes high, she visited her homeland.

There, she heard of a job teaching mathematics in a girls' high school. When she inquired about it, the Minister of Education told her the position was too inferior for someone of her importance. More likely, she was turned down for two other reasons: the first, because she was a woman; second, because as a teenager she had worked with Aniuta and other students who wished to upset the Czar's government. Sofia returned to Stockholm.

Now, almost as soon as one disappointment occurred, another followed—too fast to lessen the shock.

Her book *Struggle for Happiness* failed in Sweden. Then her beloved sister Aniuta, whom she had called her "spiritual mother," died after a long and painful illness. Sofia buried herself once more in mathematics.

Prix Bordin

Sofia decided to compete for the greatest mathematics prize of the time, the Prix Bordin, offered by the Paris Academy of Sciences. All during the summer of 1888, Sofia worked on the problem she had set herself, frequently staying up all night. The title of her research was *The Problem of the Rotation of a Solid Body About a Fixed Point*. The subject was the form of Saturn's rings.

Mathematicians try to write equations that will describe certain situations that occur in nature, such as the rotation of the earth around the sun. Some of Sofia's most important work involved the study of the shape and behavior of Saturn's rings.

Before Sofia's work scientists had considered the shape of Saturn's rings to form an ellipse. An ellipse is a special oval shape that has two axes of symmetry.

An axis of symmetry is a line across which you can imagine folding an object and fitting it perfectly on the opposite side.

Sofia proved that the cross-section of the Saturn ring must be egg-shaped. Such a shape, like a lengthwise slice of egg, is symmetric along only one axis instead of two.

Notice that if you fold the figure horizontally, the two parts are exactly the same. But if you try to fold the figure on a vertical axis, the two parts won't fit.

A Great Prize

After putting out tremendous energy on her research on Saturn's rings, Sofia submitted her work to the Paris Academy of Sciences. On Christmas Eve, 1888, she was named the winner of the great prize! Along with it, she received 5000 francs, an increase of 2000 francs over the usual amount, because her work had solved a problem that was so important to the mathematics of that time. The motto on her essay was "Say what you know, do what you must, come what may."

A Love Affair

Then, during this same year, 1888, Sofia fell in love. The man's name was Maxim Kovalevsky, and he was distantly related to her former husband's family. She called him "Fat Maxim" in tones of proud affection.

Maxim Kovalevsky. Sofia's death was a special loss, since the couple had decided to marry the following spring in Russia.

Maxim was a Russian lawyer, sociologist and historian, who was fired from the University of Moscow in Russia because he criticized Russian constitutional law. This was like criticizing the Czar himself. Neither he nor Sofia could go back to Russia and make a living.

Sofia and Maxim had many fights. She was jealous and possessive, and she went from love to anger, and back again, over and over. Maxim taught at universities mainly in France and often left Sofia. When they were apart, Sofia felt utterly abandoned. Maxim turned up faithfully when she received the Prix Bordin in Paris.

Even while she was being honored in Paris, she was very unhappy because of her love troubles.

She wrote a friend, "Letters of congratulations are pouring in from all sides, but. I am as miserable as a dog. No, I hope, for their sake, that dogs cannot be as unhappy as human creatures, especially as women."

A couple of years passed, and again it was the middle of a long, harsh winter. For the Christmas vacation, Sofia went to join Maxim at his villa in France. There, escaping from her bleak life in Stockholm, she spent many happy weeks with Maxim in the sunny atmosphere of the south of France. She wrote to her daughter that the view from their veranda showed a garden "Blooming with roses, camellias and violets, and oranges ripening on the trees."

At the end of the vacation, the couple separated and Sofia was alone on the trains going north. The weather was cold and rainy and she caught cold.

The end was swift: from cold to inflammation of the lungs to pneumonia. In three days, she was dead. She was only forty-one years old. The evening before she died she said, "I feel as if a great change has come over me." Face to face with death, she was suddenly at peace.

The brother of Gösta Mittag-Leffler wrote a poem about Sofia. Calling her the "Muse of the Heavens."

While Saturn's rings still shine,
While mortals breathe,
The world will ever remember
*   your name.*

We do remember her name today. And happily, her countrymen have honored her with a special postage stamp.

Shade all prime numbers to complete the picture of Sofia Kovalevsky.

Drawing by Analee Nunan.

Activities

• **Shade all prime numbers to complete the picture of Sofia Kovalevsky on the previous page.**

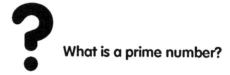

What is a prime number?

A number other than 1 is **prime** if its only factors are the number 1 and itself . Factors are divisors of the number.

examples— 7 is prime. It's only divisors are 1 and 7.
15 is not prime. It has divisors 1, 3, 5 and 15.

For this activity it is convenient to have a list of prime numbers handy. If you don't have such a list, try the following activity. It is called the Sieve of Eratosthenes, after the Greek mathematician who supposedly made up the idea.

Make the following modifications to the number grid:

- **Draw a square around ☐1 . Then remember that 1 is not a prime number.**

- **Draw a circle around ②. Two is the first prime number. Two is also the only even prime number.**

- **Now draw a horizontal line (——) through all multiples of 2. This will be all the other even numbers.**

- **Draw a circle around 3 , the next number that has not already been crossed out.**

- **Now draw a vertical line (|) through all multiples of 3.**

note— Notice the numbers crossed out with both vertical and horizontal lines. These are the **common multiples** of 2 and 3. These are also the multiples of 6.

- Draw a circle around ⑤, the next number that has not yet been crossed out. Now draw a diagonal line (╱) through all multiples of 5.

- Next draw a circle around ⑦, the next number that has not previously been crossed out.

- Draw a diagonal line going the other way (╲) through all multiples of 7. (You are just about done.)

- Go through the grid and draw a circle around every number that has not yet been crossed out. These circled numbers are all the prime numbers between 0 and 100.

1	2	3	4	5	6	7	8	9	10
11	12	13	14	15	16	17	18	19	20
21	22	23	24	25	26	27	28	29	30
31	32	33	34	35	36	37	38	39	40
41	42	43	44	45	46	47	48	49	50
51	52	53	54	55	56	57	58	59	60
61	62	63	64	65	66	67	68	69	70
71	72	73	74	75	76	77	78	79	80
81	82	83	84	85	86	87	88	89	90
91	92	93	94	95	96	97	98	99	100

MARY EVERETT BOOLE
1832-1916

"You were the roots and I the leaves of the plant ..."

The two children crouched down in the soft morning rain. The girl poked under a leaf with a stick. The boy, smaller than she, pushed aside some wild grass.

"Regardez!" the boy exclaimed, "un papillon." (Look! A butterfly.)

Its wing was torn. The girl carefully carried the creature inside the house. Taking turns, they blew on its wings to warm it.

Mary and her brother George often nursed insects that had been hurt by the frost or rain. These and an occasional lost dog were the only pets the children were allowed to have.

Their father, Thomas Everest, a minister, was seriously ill. He was under the care of a famous doctor named Samuel Hahnemann. Though Mary and George had been born in England, they moved to the small village of Poissy in France when Mary was five and George was two.

Dr. Everest's cure took six years. Life was lonely for the children in Poissy. They belonged to an English minister's household while everyone else in the town was French Catholic. Also, the Everests did not approve of French politics of that time. The laws of the French monarchy were harsh toward the people. Servants looked after Mary and George, but the two children were mostly on their own.

The Leaves and the Root

Mary was a bright child, outgoing and talkative. "You were the roots and I the leaves of the plant," she wrote to George when both were old.

Little George was his sister's fan. When Mary was nine, she wrote a play, and George, who acted in it, boasted, "C'est ma soeur qui a invente cela." (My sister made that up!)

Homeopathy

Their father was a great believer in homeopathy, a medical system promoting health and preventing disease. Dr. Hahnemann was the founder of homeopathy, and Mary's father was devoted to his system. Followers of homeopathy practiced some extreme customs, and in the Everest household, Mary and her brother George practiced them too.

They took long walks before breakfast in freezing weather and baths in ice water to help them resist disease. Some thought the cure was worse than the disease. But Mary was loyal to her father and submitted to whatever experiments he dreamed up without complaining.

Mt. Everest.

Mt. Everest and Uncle George

Mary's family name was made famous by her uncle, George Everest. George was a surveyor who spent twenty years in India. Because he led a survey team up the great mountain, Mount Everest was named for him.

Uncle George did not visit often , but when he did, he brought tales of adventure in far-off places. Mary was a great favorite of this uncle, and he wanted to adopt her, but she was too attached to her parents to agree.

A Crush

Mary had a tutor from the village. Monsieur Deplace taught Mary every morning from 6:00 to 8:00.

From the first day, Monsieur Deplace made arithmetic clear to Mary. When her mother had tried to teach her long division, it was a mess! With Monsieur Deplace, learning was easy. He asked her a series of questions. Then he told her to write down the answers. When she read them aloud to him, she realized she was reading certain orderly steps that would solve her problem. Mary never forgot this wonderful way of learning.

Mary had a crush on Monsieur Deplace. Although he was pleasant, he never showed her affection. Still, she felt she was someone special to him, and she remembered him all her life.

The Rector's Assistant

When Mary was eleven, her father regained his health at last, and the family went back to England. Her father became rector of a church at Wickwar, at the foot of the Cotswold Hills. Mary was taken out of school and became her father's assistant in his parish work. Her duties were visiting old people, teaching children in Sunday School and helping her father prepare sermons.

The Differential Calculus

For mathematically minded Mary, leaving school did not mean an end to her studies. She taught herself calculus from books she found in her father's library. While she was teaching herself differential calculus, she became stuck and she looked around for some help.

"I soon found in the library an old book of Fluxions into which I plunged with delight," she wrote. "After I had been revelling in my prize for a week, my father found me with the book and took it away, telling me that the Fluxion notation was old-fashioned and inconvenient, and quite given up now at Cambridge." Since women students were not admitted to Cambridge, Mary had no way of discovering this for herself. "I went back to my differential book, and found, to my great delight, that it was now perfectly clear to me."

How could that be? This was a puzzle.

Mathematics, Friendship and Love _____

Two years later, when Mary was eighteen, she had a chance to solve the puzzle. She visited the home of her aunt and uncle who lived in Cork, in Western Ireland. Her Uncle John was Professor of Classics at the University of Cork.

One day, a young professor, a good friend of her uncle's, came to call.

George Boole was already a well-known mathematician. Mary told George about the differential calculus puzzle and how the old-fashioned method of learning calculus had helped her. George said the book of fluxions helped her by setting up certain orderly steps for her to follow, just as Monsieur Deplace had done so long ago. Using these steps, Mary was able to solve the problems in the regular calculus textbook.

Mary liked George Boole very much. He was a kind and understanding man, as well as being a mathematical wizard. Students and children loved him. After her return to England, she wrote to him. She sent him examples of her work in mathematics. Two years later George came to Mary's home in England and began to teach her a serious course in mathematics.

At this time, George was writing a book called *Laws of Thought*. When it was published two years later, it became a sensation among mathematicians and other serious thinkers. In it, he investigated the laws that govern the part of the mind that reasons about things. These laws were expressed in an algebra of zeros and ones. This algebra even today is called "Boolean Algebra."

Later, when George revised his *Laws of Thought*, Mary, with the help of George's students, read the manuscript. When he wrote something that was not clear they told him so. Once, when he was composing a section on differential equations, Mary sent the manuscript back five times for rewriting! On the sixth time around, it was clear.

A Boolean Family _____

A few years later Mary's father died. In her grief she turned to George for comfort and friendship. A year later the two were married. Mary was twenty-three years old and George was forty.

"Into the next nine years," wrote a biographer, "were crowded a lifetime of events. Five daughters were born to them: Mary, Margaret, Alicia, Lucy and Ethel. It was a very happy marriage, and Mrs. Boole was able by her understanding care to safeguard her husband's health and protect him to a great extent from the effects of his constant and strenuous work."

Safeguarding her husband's health, as she had spent her youth safeguarding her father's health, seemed quite natural to Mary. But her life did not follow a set formula. Rather than moving along a straight line, Mary's destiny was to wind, bend and curve back and forth throughout her life.

Mary and George read many books together. They enjoyed the English poet Milton and other poets and philosophers. The two talked about new ways to educate the young, taking what

Mary's husband, George Boole, was a famous mathematician who invented a new kind of mathematics called Boolean Algebra. From *An Investigation of The Laws of Thought* by George Boole, L.L.D., Dover Publications.

was best from Monsieur Deplace, from their experience with George's students, from Mary's own study, and from other thinkers. But they made no plans for the education of their own children.

Suddenly, George caught pneumonia and died. Mary's youngest child was six months old, and she herself was only thirty-two.

What did she do? She went out and got a job!

The Wide, Wide World _____

In those days, this was a heroic task. Mary had no job experience except sermon writing parish teaching and working with George's students. But there was the key—writing and teaching.

While Mary tackled her personal problems, turmoil was also stirring up in the wider world. Revolutions had swept through Europe in 1848 and America was on the verge of

Civil War. It was 1864, and educational and political questions were boiling in England. The young widow's mind bubbled with them.

Mary had never lost the love of learning. George had been interested in the ideas of a man named Frederick Denison Maurice and wanted to invite Maurice to join their discussion circle, but George's illness and death prevented this plan.

After George died Mary pledged herself to continue George's work. When the opportunity arose she made an appointment to meet Maurice. Maurice was a lecturer at Queens College.

Queens College in Harley Street

Queens College in Harley Street, London, was founded in 1847. It was the first women's college in England. Its purpose was to train young women to become governesses. Neither women nor Jews at this time were granted college degrees anywhere in England, but at least here women could study on a college level. Its patron was Queen Victoria.

Even so, women could not receive degrees and no women could be appointed to the teaching faculty. The only jobs available were so-called "staff assistants." There was no direct teaching of students in these jobs.

When Maurice met Mary, the first thing he asked was if she would take a position as librarian at the College. Mary accepted and went to work.

Now she was in her element! Although she could not teach, she became a friend and advisor to the students. She organized and presided over what came to be called "Sunday Night Conversations."

Mary and the students discussed Boole's mathematics, Darwin's natural history, psychology, and how each subject affected the others. They held logic-practice talks.

"I thought we were being amused, not taught," a graduate wrote to her later. "But after I left, I found you have given us a power. We can think for ourselves, and find out what we want to know."

Mathematics As Fun

Mary began to teach children, using her own theories. She was most interested in showing how ordinary everyday activities prepared children to learn mathematics and science.

" ...Children do things such as drawing or sewing, counting in tens, ...sharing an apple or painting a pattern on a wall. And in the unconscious (usually not to come into consciousness for years) is growing ...(an understanding of) zero and infinity, adding or multiplying minus ...and many other fundamental mathematical (ideas)...."

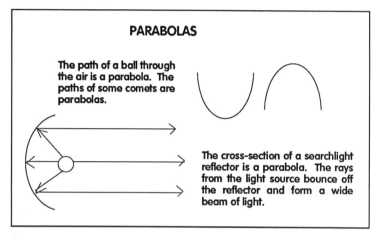

PARABOLAS

The path of a ball through the air is a parabola. The paths of some comets are parabolas.

The cross-section of a searchlight reflector is a parabola. The rays from the light source bounce off the reflector and form a wide beam of light.

Natural materials and imagination: that was the magic combination to create excitement in mathematics class. Girls in her classes used needles, thread, and cardboard to form curves with long straight stitches. Boys used their penknives to cut twigs from hedges. They took elastic from hats and slats from cigar boxes to build three-dimensional figures.

One day the Head of the London Board of Education came to Mary's class. An eleven-year-old student demonstrated a toy the class had made together. It showed a parabola in the act of changing its rate of curvature. The official was amazed at the cleverness of the students.

Children, Mary believed, should "have the opportunity of watching how one geometrical type-form grows out of, or flows into, another. A common night-light placed in the bottom of a deep round jar in a dark room throws on a sheet of cardboard held over it patterns of conic sections, which pass into each other as you change the position of the cardboard." Children love to watch the shadows change and it is good training for geometry, she advised.

Mary threw all these ideas together into a pot and out came a rich stew—a book called *The Preparation of the Child for Science* .

Curve Stitching

Mary invented cards marked for the purpose of curve stitching. They were known as "Boole cards" in England. Mary happened upon curve stitching, or what today we call string geometry, by chance, and saw at once it could be an aid in learning about the geometry of angles and space.

"In my young days," she wrote, "cards of different shapes were sold in pairs, in fancy shops, for making needle-books and pin cushions. The cards were intended to be painted on; and there was a row of holes around the edge by which twin cards were to be sewn together. As I could not paint, it got itself somehow suggested to me that I might decorate the cards by lacing silk threads across the blank spaces by means of the holes. When I was tired of so lacing that the threads crossed

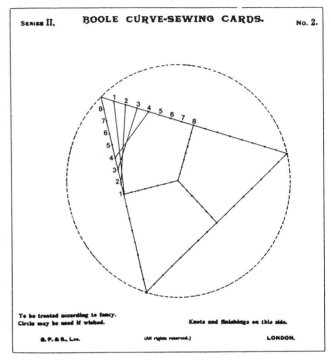

in the centre and covered the whole card, it occurred to me to vary the amusement by passing the thread from each hole to one not exactly opposite to it, thus leaving a space in the middle. I can feel now the delight with which I discovered that the little blank space so left in the middle of the card was bounded by a symmetrical curve made up of a tiny bit of each of my straight silk lines; that its shape depended upon, without being the same as, the outline of the card . . ."

A book about experiments with curve stitching was written by a friend of Mary's. It was called *A Rhythmic Approach to Mathematics*. Cards with patterns from ancient times appeared in the book.

"Some of the patterns reproduced designs in old Celtic art; others in old Egyptian and Greek art. In fact," Mary wrote, "we are hearing from various parts of the world. (People say) `you have reproduced the ornamental work on such or such a very old building.'"

The designs reproduced basic ones that were used in Egyptian, Greek and Celtic art! To Mary, this showed that people's unconscious minds were similar no matter when or where they lived.

"Psychic" Science

In Mary's time, many people asked questions about the spirit world. They wondered whether it existed and, if so, how it worked. They used the word "psychic" for anything to do with the spirit world. Mary had long thought about psychic happenings. She completed a book and called it *The Message of Psychic Science for Mothers and Nurses.*

At this time, Maurice was her employer, as well as a minister in the Church of England. Maurice prided himself on being a tolerant man, but he stopped short of a book about psychic science. Mary's book not only talked about physical health, but also mental health and called it science. Mary's book maintained there was such a thing as thought transference between people. This was going too far, Maurice thought!

Friends of his who also belonged to the Church of England blocked the publishing of Mary's book. It took fifteen long years before the book was finally printed. By that time Mary was no longer in Maurice's employ, and his opposition could not keep her from publishing.

A Fair Exchange

Mary went to work as secretary to James Hinton, an old friend of her father's. Hinton wrote about evolution and also about the art of thinking. It was the second subject that drew Mary to work with him.

Each had something to give the other. They studied the development of the mind. Mary taught Hinton mathematics and how to apply equations to the art of thinking.

The Magic of Numbers

It is possible, Mary believed, to express the basic notions of the universe in equations made of numbers and symbols. The number "1," for example, is the expression of unity in the universe. Zero is infinity. Anything could be translated into mathematical symbols: a rainbow, a butterfly chrysalis, a dust spiral. Mathematics gave power to discover truths in all fields of knowledge.

Mary wrote and talked about her beliefs with mystics from the East and psychologists from America. Founders of the progressive education movement in America studied her writings.

The Cranks

Mary had a group of friends who called themselves "the cranks." She had met them at a vegetarian restaurant in London. After a time, they put out a magazine called "The Crank." Mary contributed many articles with such catchy titles as "Are We Berserks or Christians?"

For the next thirty years, Mary wrote many books and articles. Her titles show the wide range of her interests: *Mathematics in Occultism*, *The Divining Rod*, *The Schoolgirl Medium*, *About Girls*, *What One Might Say to a Schoolboy*, *Hooliganism* (juvenile delinquency), *Philosophy and Fun of Algebra*, *The Logic of Love*.

For the twentieth-century reader, leafing through Mary's Collected Writings is a little like taking a trip with Alice through the

Mary Boole at her writing desk. She wrote about many different subjects during her long life.

looking glass. All sorts of odd ideas come skipping along, such as, "It is the moral duty of people not to go insane." " Geniuses should live to be old rather than `burning out' and dying young." " Having secrets is a mistake."

Mary made a rule for herself, when she was thirteen, to never to keep secrets. That way she could publish whatever she wished and say whatever she chose.

Except for Lucy, who lived with her, all of Mary's daughters were grown now and on their own. Lucy was a chemist and she lectured at the London School of Medicine for Women. Mary's daughter, Alicia, was a mathematician of considerable talent who was said to have the ability to visualize figures in a fourth dimension. The youngest, Ethel, was a novelist.

Mary spoke at various clubs and societies, such as the Parents' National Educational Union and the Christo-Theosophical Society. These meetings were open to all, and all kinds of opinions were welcomed. They were attended by foreign scholars, ministers, editors, educators and other thinkers.

A Shocking Dream

Mary had grown up on religious sermons and found it hard to resist what she called "a little bit of preach." Yet she was always ready to deflate windbags or people she called "prigs." She may have been a bit of a prig, but she didn't mind telling stories on herself.

For example, Mary had a dream which kept repeating itself.

In the dream, "I find myself in the street without proper clothing, without bonnet and cloak, or even my nightdress."

When she had this dream, she always looked over the manuscripts which were waiting to be taken to the printer. She wanted to be certain that she had not revealed too much of herself in her writing. She did not want her "naked" feelings to show to the world.

World War I: The End of an Era

When World War I started, Mary's health was failing. She was sorry she was too weak to knit sweaters and blankets for the war effort. Her contribution was to open her house to the many people who knew her.

"They came and found a quiet place for an hour, away from the turmoil of a country at war and the terrible news in the newspapers."

Mary was quite old by now. She had breathed life into wounded insects. She had given birth to five children, all living. She had given the power of thinking to students so they could find out what they needed to know to live well. She had enlivened mathematics classes for countless girls and boys.

Mary had been widowed for fifty years. She had raised her family alone and now relatives, friends and even her eyesight were slipping away.

"Ah me!" she wrote, "But it is a lonely world!" But she had a strong religious faith, and so even sadness could be woven into a rich cloth. "If one knows the artistic way of using the world's shadows, discords and lurid dark silks."

Mary died at the age of 84. Her life spanned the Industrial Revolution in England. This period brought to England a new set of political and social challenges, including public education, public health, trade unions, the cooperative movement.

Mary called herself a mathematical psychologist. This meant she tried to understand how people, and especially children, learned mathematics and science, using the reasoning parts of their minds, their physical bodies, and their unconscious processes.

Her work influenced many others of her time. "Within the first decade of the present century," wrote her biographer, "new methods of teaching many subjects had been developed. Experimental work went on."

Today, in your own classroom, you may learn subjects the way Mary Boole taught them.

Activities

| • Curve stitching |

Mary Boole, developed a technique for making attractive designs by threading colored string through cardboard forms. Pictures of two of these cards, called Boole Curve-Sewing Cards, are included here. **Curve stitching shows an interesting mathematical idea. Curved lines can be formed solely by a series of straight lines.** How can this amazing thing happen? Let's try it and see.

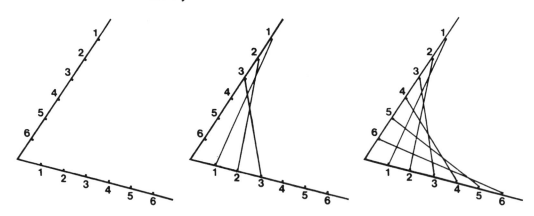

Notice the numbers drawn on both rays of the angles above. Draw a line connecting point 1 to point 1, another from point 2 to point 2, a third from point 3 to point 3. Keep doing this. Watch the trace of a curve appear as the lines are added.

If the spacing between points is kept the same while the width of the angle is changed, the shape of the curve will change. **Connect the points in the following figures and see what happens**

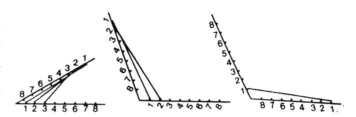

Notice what happens when the number of points on each edge is increased while the angle is kept the same.

The more points, the smoother the curve!

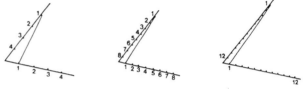

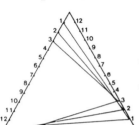

Try connecting points on edges of other geometric figures. For example, try connecting the points on this triangle.

Experiment with other designs.

Make your own numbered shapes. Try coloring your designs like the illustration below. Many attractive designs can be made this way.

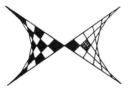

Here and on page 52 are pictures of two of Mary Boole's cards with numbers added in a few places. Connect the points in the usual way and watch the curved designs appear. Try connecting different edges. See how the designs change.

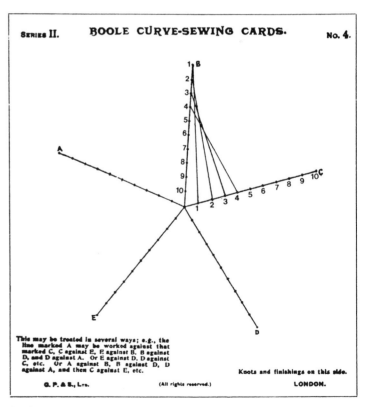

LINE DESIGNS by Dale Seymour, Linda Silvey and Joyce Snider is a book which suggests many more ideas you can try.

• Square numbers

There are several way to think of square numbers.

• Multiply a number by itself. (Use a calculator if you like.)

Fill in the blanks to see square numbers with three or less digits:

1 X 1 = _____   11 X 11 = _____   21 X 21 = _____

2 X 2 = _____   12 X 12 = _____   22 X 22 = _____

3 X 3 = _____   13 X 13 = _____   23 X 23 = _____

4 X 4 = _____   14 X 14 = _____   24 X 24 = _____

5 X 5 = _____   15 X 15 = _____   25 X 25 = _____

6 X 6 = _____   16 X 16 = _____   26 X 26 = _____

7 X 7 = _____   17 X 17 = _____   27 X 27 = _____

8 X 8 = _____   18 X 18 = _____   28 X 28 = _____

9 X 9 = _____   19 X 19 = _____   29 X 29 = _____

10 X10 = _____   20 X 20 = _____   30 X 30 = _____

• Analyze another pattern of dots. Add the next odd number.
  Notice how the pattern grows.

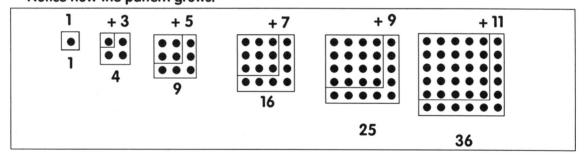

• Analyze another pattern of dots. Add adjacent triangular numbers.
  Notice how the pattern grows.

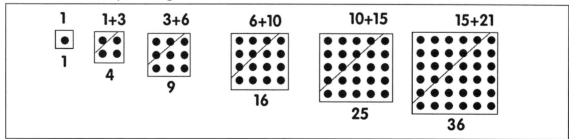

(See the activities section in *Ada Lovelace* for more information about triangular numbers.)

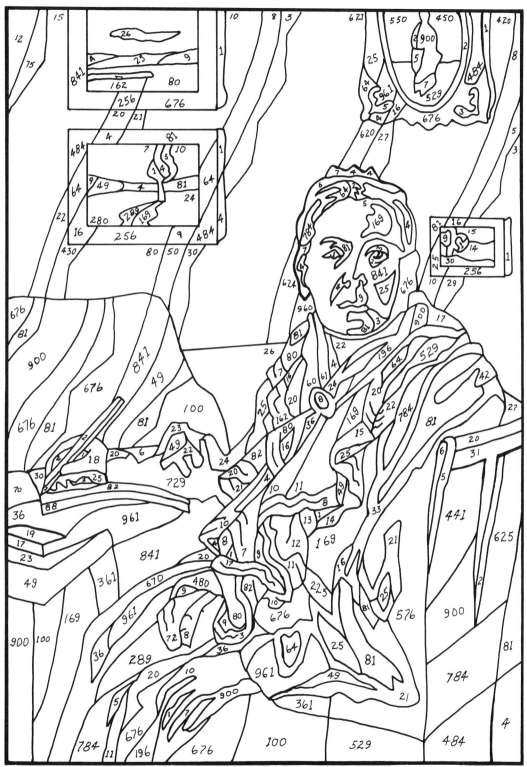

Shade all square numbers to enhance this drawing of Mary Boole.
Drawing by Analee Nunan.

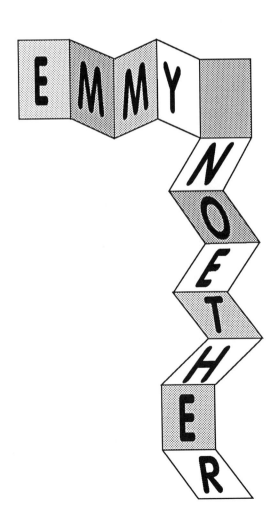

EMMY NOETHER
1882-1935

"...the greatest of women mathematicians,
a great scientist, an amazing teacher,
and an unforgettable person..."

An Ordinary Childhood

When Emmy Noether was a young girl, no one would have predicted greatness for her.

Emmy was born in 1882, the oldest child in her family. She had three brothers, but two of them died at a young age. Her surviving brother, Fritz, was two years younger than Emmy.

The Noether family were Jewish and lived in Erlangen, Germany. They valued learning very highly.

Emmy's father, Max, was a mathematics professor and research scientist at the University of Erlangen. He was a kind, jolly sort of man. Her mother, whose name was Ida Amalie, was a typical "hausfrau" and kept a tidy, well-run household.

For her entire childhood, Emmy lived in the same apartment in Erlangen. She was very near-sighted and wore thick glasses. She learned the skills that were taught to girls of the middle class. She learned to play the clavier, which is something like the piano. She cooked and dusted. She liked going to family parties where there was dancing.

Emmy Noether 1882-1935. Reprinted with permission of Birkháuser Boston Reprinted from *Emmy Noether,* edited by James W. Brewer & Martha K. Smith, 1981; by courtesy of Marcel Dekker Inc..

Emmy was clever as well as friendly and cheerful. From the start, she liked languages and studied French and English. When she finished high school she took a test and was certified to teach French and English in schools for girls. If she had to, she could now earn her living. Life was following the expected pattern for a young woman of her time and place.

But Emmy's was not an ordinary family, and times were changing. German universities were beginning to admit young women to earn degrees. Not only was her father a mathematician but her brother Fritz was following in his footsteps. Fritz entered the University of Erlangen to study mathematics. Perhaps this is why Emmy, as an eighteen year old, began to sit in on classes at the University. She did this for two years. Later she took the examinations for entrance as a doctoral student in mathematics and passed. Now she was a student in good standing and a different kind of career was opening for her.

Five years later, Emmy received her doctoral degree. She was the second woman in the history of the University to receive a doctoral degree in mathematics. The first woman had received her degree the year before Emmy.

A Degree But No Job

Now Emmy Noether was really ready for her big career. There was only one problem No women were permitted to take jobs as mathematics professors in Erlangen, or indeed, anywhere in Germany!

For the next ten years, from 1909 to 1919, Emmy worked without pay at the Mathematics Institute in Erlangen. Partly, she helped her father in his work and partly she worked on her own research. Sometimes she taught in her father's place when he was sick. Slowly, her reputation grew. She began to publish papers about her work. In her lifetime she published forty-five papers.

The first World War began in 1914. Emmy was a pacifist. She had high hopes that this terrible war would end quickly and there would be a lasting peace in Germany. In 1918, Germany lost the war. The German monarchy was overthrown, the country became a republic, and women were given the vote. But even though Emmy could vote, she still received no salary whatsoever for her work.

A Growing Reputation

Two great mathematicians, Felix Klein and David Hilbert, were working at the University of Göttingen in Germany. They heard of Emmy and thought she could help them in their work on Einstein's general theory of relativity. They invited her to move to Göttingen in the hope that she could join the faculty.

But there were no women on the Göttingen faculty. Other faculty members said, "What will our soldiers think when they return to the University and find they are expected to learn at the feet of a woman?"

Throughout these arguments between Hilbert and the rest of the faculty at Göttingen Emmy remained serene. She did not become bitter. She thought people were good, and kept that belief uppermost in her mind. Emmy was now thirty-seven years old, and her reputation as a brilliant mathematician was growing fast.

Finally, she received a place as a lecturer at the University of Göttingen. Now, instead of teaching for other professors, she could teach courses under her own name. But she was still not getting paid. Not until three years later did she finally receive a small salary. This came just at the right time because Germany's post-war economy was wracked by inflation, and money from her family's estate, which had helped to support her, was running out.

The Noether Boys

Emmy gathered a group of students around her at Göttingen. Her growing success as a mathematician attracted outstanding students. They came from Russia, the Netherlands, and other countries. They made a lively group these "Noether Boys." Years later, when Emmy taught at a women's college in the United States, her group was known as the "Noether Girls."

Emmy's style of teaching was very challenging. Her explanations were, as her student Van der Weyden put it, "rattled off at top speed." She would launch into explanations, feverishly writing symbols on the board, leaving unfinished sentences in her wake, and a group of puzzled students sitting in front of her. Perhaps, being so near-sighted, she never noticed their expressions.

Once, after she had given a lecture to a large group of students—some regular students and some visitors—one passed a note up to her. It read, "The visitors have understood the lecture just as well as any of the regular students."

But not all of her students were confused. A few patient and attentive ones caught on and became loyal followers. They learned to snatch the ideas she flung out and put them to use in their own research. In this way, they advanced and became distinguished mathematicians in their own right. They never forgot their debt to Emmy Noether, and some gave her credit in their own works.

One of her former students, Alexandroff, invited her to Russia to the University of Moscow. He later wrote about her visit: "Emmy Noether very easily fit herself in with our life. She lived in a modest room in the KSU hostel near the Crimean Bridge, and most of the time she walked to the University. She was very much interested in the life of our country, especially in the life of Soviet young people."

Sharing ideas made attending professional meetings exciting for mathematicians and other scientists. Noether, with colleagues, at a theory conference in 1921. From *Mathematical People: Profiles & Interviews* edited by Donald J. Albers and G.L. Alexanderson, 1985.

All during the twenties, Emmy's work in algebra progressed. Her work was part of the new algebra. Her work did not rely on adding or multiplying numbers or solving equations. Instead, her work in algebra dealt with ideas.

A Brilliant Mind, A Warm Personality

Emmy was generous with her genius. She never hoarded her brilliant ideas. Her students were like her own family to her. She was interested in their personal lives, and listened to their problems. She was "warm, like a fresh loaf of bread," said Herman Weyl, one of her biographers.

Emmy had a deep, loud voice. She laughed heartily. She was short and heavy-set. She looked solid, earth-bound. Yet her ideas soared with ease and grace, as did her lively spirit.

Emmy Noether surrounded by colleagues, near Göttingen, 1932. Left to right: Ernst Witt, Paul Bernays, Helene Weyl, Hermann Weyl, Joachin Weyl, Emil Artin, Emmy Noether, Ernst Knauf, unidentified person, Chiugtze Tsen, Erna Bannow Reprinted from Emmy Noether, edited by James W. Brewer & Martha K. Smith, 1981; by courtesy of Marcel Dekker Inc..

Departing the Homeland

In 1933 the Nazis, who had come to power in Germany with Hitler at their head, demanded that Jews be thrown out of all university positions. Emmy's brother, Fritz, had to move with his family to Siberia where he was offered a position at the University of Tomsk.

Emmy too had to leave Göttingen and the beloved country of her birth. Alexandroff desperately tried to secure a place for her at the University of Moscow. But before he could do this, American friends found her a job as visiting professor at Bryn Mawr, a women's college near Philadelphia. Emmy moved there in 1933. She was fifty-one years old.

A Woman Mathematician in America

Anna Pell Wheeler was Head of the Mathematics Department at Bryn Mawr when Emmy arrived. She had been instrumental in bringing Emmy to Bryn Mawr. Professor Wheeler had studied at the University of Göttingen. She could understand how Emmy's career had been blocked in Germany because of her sex. She could understand the shock of being uprooted from Germany and transplanted to another world.

This was the first time Emmy had a department head who was both a mathematician and a woman. Up to this time, all her colleagues had been men. When Emmy's old friends and former students came to Bryn Mawr to visit her, she introduced Professor Wheeler as her good friend.

As a professor at Bryn Mawr, Emmy Noether made quite an impression. Betty Morrow Bacon, a student at Bryn Mawr at that time, recalls meeting Emmy soon after her arrival in America.

"Before she came we were told, `This is one of the great people alive in the world today. It is an honor to have her at Bryn Mawr.'

"When I first saw her striding across the campus, I felt a little scared. Her expression seemed stern and forbidding. She was large and solid-looking and she wore a long dress. She didn't speak. She was not at all like other people I knew.

"There was a faraway look on her face, as though her mind was not in this world at all. What was she thinking of? Maybe she missed her old life in Germany. Maybe higher mathematics filled her thoughts. I never knew."

Another student at Bryn Mawr, a graduate student in mathematics, Grace Shover Quinn, recalls her impressions of Emmy after she was settled into her new life at Bryn Mawr.

Emmy Noether "was around five feet four inches tall and slightly rotund in build. She had a way of turning her head aside and looking into the distance when trying to think while talking. Her lectures were delivered in broken English. She often lapsed into her native German when she was bothered by some idea in lecturing.

One of the last potographs of Emmy Noether taken at Bryn Mawr in April, 1935. From Mathematical People: Profiles & Interviews edited by Donald J. Albers and G.L. Alexanderson, 1985.

"She loved to walk. She would take her students off for a jaunt on a Saturday afternoon. On these trips she would become so absorbed in her conversation on mathematics that she would forget about the traffic and her students would need to protect her."

Another young woman, who became a prominent mathematician herself, remembers Emmy Noether at both Göttingen and Bryn Mawr. In her reminiscences, Olga Taussky-Todd writes about her own arrival as a graduate student at Bryn Mawr, fresh from abroad. She, like Noether, knew only minimal English and was quite alone. Emmy Noether was just beginning her second year at Bryn Mawr, and Olga was one of four women graduate students who ended up working with her.

Olga had known Emmy Noether at Göttingen. "She was definitely popular with the students, I had the good fortune to gain her confidence, and we became good friends."

Besides teaching and doing research at Bryn Mawr College, Noether gave weekly lectures at the nearby Institute for Advanced Study in Princeton, New Jersey, where Albert Einstein and other famous German refugees now worked.

Olga writes about accompanying Emmy Noether to her Princeton seminars. "On her weekly trips to Princeton I accompanied her frequently, though not every week because of the high train fare. She was very pleased that I went with her, and we had nice chats." These trips to Princeton were the highlight of her year in Bryn Mawr. "Since I travelled with Emmy I was invited to dinners in the evening together with her. The Institute for Advanced Study and the department of mathematics were in the same building; I could see people like Einstein walking in the corridor, and I was even introduced to him several times."

Olga Taussky-Todd, 1935, dressed for Bryn Mawr College Commencement. Taussky-Todd has become a prominent mathematician. From Mathematical People: Profiles & Interviews edited by Donald J. Albers and G.L. Alexanderson, 1985.

When Olga Taussky-Todd returned to England the following year she looked back with nostalgia at her Bryn Mawr days. Although she found Cambridge University wonderful, she couldn't help remembering her Bryn Mawr year with its rich female mathematical contacts "the Noether girls", Mrs. Wheeler, the Princeton excursions, and Emmy Noether herself. It would take a long while for her to replace these connections.

Death: A Great Shock

In 1935, Emmy Noether entered the hospital to have an operation. Taussky-Todd writes that Emmy was only fifty-four years old at the time.

"Unbeknown to all of us, she was ill! She tried to hide that fact. But when certain troubles bothered her increasingly, she confided in a doctor at Bryn Mawr and he persuaded her to undergo surgery immediately. A week later she died of heart trouble. At least this is what we were told."

Emmy Noether's death was a great shock to her friends all over the world. Soon after her death her Russian friend, Alexandroff gave a talk to the Moscow Mathematical Society. He said, "Emmy Noether was the greatest of women mathematicians, a great scientist, an amazing teacher, and an unforgettable person."

Abstract Algebra

Emmy Noether worked in a part of mathematics that is called abstract algebra. This kind of algebra is quite different from the algebra you learn in school.

In abstract algebra people talk about GROUPS and RINGS and FIELDS. Fields are more complicated than rings, and rings are more complicated than groups. Believe it or not, the simplest arithmetic you learn in school is far more complicated than many examples of groups or fields or rings.

If you don't believe this, try the examples in the activities section.

Activities

Meet Sticky

Imagine a stick figure that can move only its two arms . . . and these two arms can only move up or down.

Suppose we want to describe the motions Sticky can do. If we do so in a certain way, we can turn Sticky and Sticky's motions into a full-fledged mathematical animal called a group.

Let's think about the four possible ways that Sticky can move, and label these four ways 0, 1, 2, and 3.

This is MOTION 0.
Here Sticky does not move at all.

This is MOTION 1.
Here Sticky moves right arm once.*

This is MOTION 2.
Here Sticky moves left arm once.

This is MOTION 3.
Here Sticky moves both arms once.

If Sticky moves right arm twice, Sticky will be back in first position.

So we can write an EQUATION.

---

* By "move once" we mean, if arm(s) is straight out it moves UP. If arm(s) is UP, it moves STRAIGHT OUT.

Just as we have learned in ordinary arithmetic that 3+3=6, we can say here:

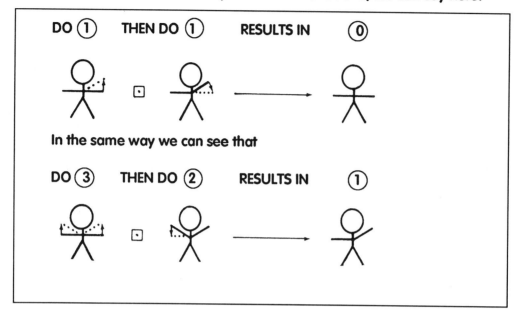

In the same way we can see that

See if you can fill in the table on the right. For each square first do the motion that the figure at the top is doing; then do the motion that the figure at the side is doing.

 How do you end up?

Look at squares that are filled in in order to make sure you understand.

? When you fill in all squares, do you see that the figures make a pattern?

Clock Arithmetic Activities _____

Another example of a group is the special arithmetic called clock arithmetic.

• Think about the face of a round old-fashioned clock.

?

● **How many numbers do you see on it?**

Suppose it is 10:00. You tell your friend you will be back in 3 hours. What time will you be back? Right you are. You will be back at 1:00.

You can write a funny-looking arithmetic sentence . . . 10 (+) 3 = 1

Notice the circle around the +. That tells those who know, that we're doing addition around a circle. This is the special clock arithmetic. Look at the clock and finish the following sentences.

$$5 \oplus 8 = \underline{\hspace{1.5cm}}$$

$$8 \oplus 5 = \underline{\hspace{1.5cm}}$$

$$11 \oplus 4 = \underline{\hspace{1.5cm}}$$

Now in ordinary arithmetic   3 + 2 = 5

$$1 + 4 = 5$$

$$2 + 3 = 5$$

$$4 + 1 = 5$$

$$0 + 5 = 5$$

$$5 + 0 = 5$$

These are all the different ways you can make 5 by adding just two whole numbers.

How about clock arithmetic?

? In clock arithmetic, how many ways can you make 5?

A You can say that   11 ⊕ 6 = 5

10 ⊕ 7 = 5

I'm sure you can think of many other ways.

Now let us consider a very simple example of clock arithmetic and its solution in the square on the next page. This clock has only 4 positions that matter. We can think of these as follow:

 0 means stay where you are.

 1 means move head 15 minutes.

 2 means move ahead 30 minutes.

 3 means move ahead 45 minutes.

Now do the additions as you did before, and complete the squares.

• Remember, 1 (+)1 = 2 means move one stepclockwise. Then move one more step clockwise, and you've arrived at position

2. How about 2 (+) 2 = _____ ?

• Notice the pattern of numbers when you've completed the square.

• Compare the difference in the patterns of the clock square and the Sticky square.

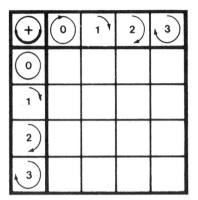

The pattern in each square comes out of exactly 4 numbers (or elements) and a rule for combining them.

?
• Why are the solution patterns different for these two squares?

Solutions _____

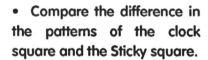

• Notice that for both squares, each number appears only once per row and column.

• If you were to fold the large square along the northwest/southeast diagonal, each half would match. This diagonal is a line of mirror symmetry.

• Notice how the number patterns differ for each square.

Emmy Noether's work gave mathematicians new tools to solve old problems. In particular, she developed important new ways of classifying these kinds of situations in rings, the next more complicated structure after groups.

Common Multiples of 2 and 3

• All numbers that are common multiples of 2 and 3 are numbers that are multiples of 2 as well as multiples of 3.

• All multiples of 2 are even numbers. Therefore common multiples of 2 and 3 are even multiples of 3.

• See the activities section in Lenore Blum, titled *multiples of 3* test.

Using common multiples of 2 and 3, shade in Emmy Noether's picture.

Drawing by Analee Nunan.

LENORE BLUM
1942-

"To see a World in a grain of sand,
And a Heaven in a wild flower;
Hold Infinity in the palm of your hand,
And Eternity in an hour." —William Blake

A Tropical Paradise

The airplane carrying the two girls descended, and the city of Caracas, Venezuela, rose up to meet them. It was the dry season and fires in the mountains caught their eye. They saw small aircraft dropping chemicals to put out the flames. As they approached the airfield, the sun illuminated a landscape of palm trees and flowers, modern buildings and streets. To Lenore, the older one, this was paradise, and it was her new home!

Lenore and her sister had left their home in New York. The whole family was coming to live in South America. Here, Lenore's father hoped to make a good living in the import-export business.

The drive from the Caracas airport to their new home was full of strange sights. An attempted revolution was taking place. Opposition forces, who wanted to overthrow the dictator, had thrown tacks in the streets to puncture tires and disrupt the city. Christmas trees were tied to front fenders of cars to sweep away the tacks.

Their lodgings were wonderful—several rooms rented in a large house. Views from second-story windows showed the lush mountains. For the first time in her life she had a back yard. Unknown in New York! In the yard, a parrot squawked—gorgeous, brilliant-hued, tropical.

Lenore was nine and her sister, Harriet, was seven. They were going to attend the local school where only Spanish was spoken and no foreigners were enrolled. This would be quite a change for Lenore. Most of her friends had been American in the old school. Many were Jewish as she was; in fact, in New York she had hardly known anyone who was not Jewish.

65th Avenue, Queens

Home in New York had been an apartment on 65th Avenue in Queens. Her earliest memory was standing at the window watching her father, Irving, when he arrived home after World War II. From the candy store on the corner, she bought *The Forward*, a Yiddish newspaper, for her grandmother, who lived with the family. Her grandmother

Mother, father and one and a half year old Lenore, in Queens. Dad is home on leave.

spoke only Yiddish. Lenore understood Yiddish, and although she didn't speak it herself, she translated her grandmother's words into English when people came to visit.

When she was young, relatives commented to each other, "How artistic Lenore is!" Art was her favorite activity, though she usually didn't show her projects to others.

Lenore visited libraries and museums with relatives and even by herself. She loved the dinosaurs at the Museum of Natural History and the wild animal habitats. Sometimes she traveled by subway to the New York Public Library and the great museums in Manhattan—the Metropolitan Museum of Art and the Museum of Modern Art. They filled her with wonder.

Her cousins, Ellen and Shelley, lived in her neighborhood and they, Harriet and Lenore played together. This was in the early days of television and on one program, called The Magic Cottage, children applied through the mail to be on the show and win prizes. All four girls eagerly applied. Shelley and Harriet were chosen. Ellen and Lenore talked and persuaded, persuaded and talked until the other two agreed to give up their places. Ellen and Lenore appeared on the show.

Many of the prizes were of some value, such as a child's record player. But Lenore saw just what she wanted: a puppet which folded up into its own little case!

"Why didn't you pick the record player? It's much more valuable," the grownups wanted to know. Valuable? To the others, maybe, but not to Lenore.

Growing Up

The women in Lenore's family were very proud of their professions and she grew up with the idea that a profession was a very important thing for a woman to have. Rose,

Lenore's mother, taught science in a New York high school. At one time she had studied to become a doctor. Rose's sisters often came to the house to visit Lenore's grandmother. Two were lawyers and a third was a teacher.

They often spoke of their older brother, Sol, who had become a medical doctor. Though he had died at the age of 32, his presence was strongly felt. As the oldest child of an immigrant family, he had led the way.

Every now and then Lenore would take a violin from the hall closet that belonged to her Uncle Sol. She was like this uncle; she would do something special. It didn't really matter to her that later, when she took violin lessons, she showed no talent for the instrument. What mattered was that she felt there was a tradition of scholarship for her to follow.

Lenore went to a progressive school which didn't give grades. She loved school and loved the summers at Far Rockaway beach where she played with her sister and her cousins. She used New York, the largest American city, as her playground. She had plenty of free time. No one kept close track of her. This suited her perfectly.

Suddenly, she was moving away to South America.

There was one flaw, a fear that came and went: her beloved grandma might become ill when no one else was there. But the years of her childhood passed and her fear was never realized.

South American Scenes

Lenore's first glimpse of Caracas confirmed her belief that good things would always happen to her.

The first day of school shattered the perfect picture. The boys in the class had learned a few words of English from American movies. "I love you," they whispered as they ran around the two American girls. How humiliating!

Then there was the principal. Here were two young American "niñas" who "no hablan Español." But they must learn to speak Spanish—pronto! The principal would teach them one hour each day in his office, beginning with the subjunctive tense of the verb. Disaster! Lenore hadn't learned grammar even in her native English.

A further humiliation: school officials put Lenore back a grade so that her age would match her classmates. In the classroom the teacher, Señorita Candelaria, used a method of teaching that was altogether strange to Lenore. The students were given blue "cuadernos" or notebooks. They had to copy stories and pictures into their notebooks over and over until they matched those in their textbooks. Every cuaderno in Lenore's class looked exactly alike!

Meanwhile, right outside the windows, lay a city which at that time was one of the most beautiful in the world. Such a bright world, too, after sunless days in New York.

Two weeks was all Lenore lasted. "I quit!" she announced to her parents. And that went for her sister, too.

A Year of Freedom

Lenore's mother, a veteran teacher, wasn't too upset. She would educate her daughters by exploring this brand new place. So, for one year, Lenore, going on ten years old, and her sister, seven, did not go to school at all. Instead, they traveled on buses each day with their mother.

They took in all the sights of the city. No formal lessons now, only informal ones that the city and its inhabitants had to teach. At the end of the day they went to the cafe where the "foreigners" met. The girls ordered ice cream sodas. Life became much easier for Lenore.

Homesickness

Even so, homesickness crept in. Lenore expressed it by writing a poem. It spoke about all the things she had left behind, about Thanksgiving, about her grandma. It was filled with longing. Her poem won a prize and was published in the English language newspaper, the Caracas Daily Journal.

She also made a pencil sketch of her grandmother. The background for the sketch was engraved in Lenore's memory: the comfortable cushioned chair her grandmother had sat in, the gooseneck lamp so the elderly woman could sew and read the newspaper, the wall paper with its pink and rose flowers, the photograph of her grandmother's parents—Russian peasants in kerchief and cap—mounted on the wall, the piano bench on one side of the chair and the little table for her sewing on the other.

Lenore's great grandparents in Russia, at the turn of the century.

But try as she might, Lenore couldn't get her grandmother's face right, not the way it really was. She left the sketch unfinished.

A new year started. A decision would have to be made about the girls' education. Seventy thousand Americans lived in Venezuela. They were mostly oil company employees and their families. Many were wealthy and they had their own schools. The one in Caracas was called Escuela Campo Alegre.

Escuela Campo Alegre — The American School

Lenore's family could not afford the tuition at this school so Lenore's mother went back to work, teaching at the American school. Lenore and Harriet began classes there.

Long division was being taught. The teacher gave an explanation. Lenore understood instantly. Though she had fallen behind because of the year's absence, she caught up and moved to the top of her class. From then on, until she graduated from high school at the age of sixteen, Lenore was the top student and mathematics was her favorite subject. Although she was a girl and the best student in mathematics, the others did not look down on her. They simply thought she was "different." Lenore made plenty of friends.

An Impression

In time her family met another Jewish family in Caracas. There were four sons in the family. The oldest, whose name was Manuel, was serious and a little shy. From their first meeting Lenore made room in her heart for this boy. He was like her; they were simpatico. He was the one she would one day marry.

Lenore rarely saw Manuel in the next few years. He was four years older than she and soon went off to the States to study at the Massachusetts Institute of Technology (MIT). But he had made an impression.

Los Pavos

Like many teenagers, Lenore had many images of how she wanted to be. She was trying out a variety of roles and sometimes her actions seemed contradictory, even to herself.

When she was in high school, her friends were "los pavos," which was Caracas slang meaning the cool crowd. One day, while she was riding with a boyfriend on his motorcycle, they had an accident and crashed. Lenore fell on her head. She felt the wound where blood was clotting.

"My brains! My brains are spilling out!" thought the panicked thirteen-year-old. But, with only a head patch and two weeks' time, she was well again. Then and there she decided that motorcycles were not worth the risk.

Cars, clothes, bikes and romances were part of the American school scene. Money was thrown around and parties were frequent. Lenore's parents were tolerant. To them, she was the older daughter, the little girl who had taken care of her grandmother so faithfully. They set no limits, so Lenore set her own: a 1:30 curfew, and she never came home later.

A Coup

When Lenore first arrived in Caracas, an attempted revolution was taking place. Five years later, on New Years Day in 1958, the dictator, Colonel Marcos Perez Jimenez, was finally overthrown. The citizens of Caracas were jubilant. The hated secret police fled and the people ran the country. Boy Scouts directed traffic and university students helped maintain order. Sometimes students who were friends of Lenore's took her with them on their patrols after the curfew. She loved being part of the excitement. The controlling junta (the new ruling group) promised elections in the fall. Every five years thereafter, free elections have taken place in Venezuela.

Searching for a Profession

Mathematics was Lenore's favorite subject. Instruction in mathematics at the high school Lenore attended was poor, and so she learned a whole year of mathematics on her own. When she expressed an interest in going on with mathematics in college, her teacher advised against it.

"Everything important was discovered 2000 years ago," he told her. "You don't want to go into a dead field."

Was that true? The sixteen-year-old transplanted New Yorker did not know. All she knew was when she looked at a proof on the board, her heart swelled. "It's so beautiful, so perfect." Mathematics cared for no one's opinion.

Buildings, beautiful modern concrete forms were springing up all over the city of Caracas. Architecture was exciting, creative. Here was a career field that combined her two loves—mathematics and art. That is what her college major would be.

But where to apply? To MIT, of course, where Manuel was and where so much exciting work was being done.

To Lenore, MIT was the pinnacle, but, to her great dismay, she was not admitted. In response to her application, MIT officials said, "We have only twenty beds in the girls' dormitories and incoming students must live on campus." Only very few women were accepted. [This excuse of dormitory space was corrected some years later. One of the first women to graduate from MIT, Katharine McCormick, gave a large sum of money to be used solely to build a dormitory for women students.] Lenore decided to go to Carnegie Institute of Technology in Pittsburgh, Pennsylvania, instead.

Graduation Summer

Lenore was valedictorian of her high school graduating class. Manuel came to Caracas for the ceremony, and her senior prom was their first date. That summer, the two young people talked about such things as Freud and modern psychology, philosophy, and the make-up of the brain. A neuron could be expressed mathematically. Fascinating! More than ever now, she looked forward to college.

Lenore and Manuel on one of their first dates (1960).

From Caracas to Pittsburgh

Pittsburgh was an industrial town, a maker of steel—old and dingy. Quite the opposite of sparkling Caracas. The students of architecture at Carnegie Tech were very serious about their career goals. They worked together in a room that took up an entire floor of a building. Often they worked through the night to finish projects. Older students offered criticism and advice. Everyone was helpful and friendly.

Mathematics is an important tool for architects and they were interested in learning formulas. Lenore, on the other hand, was interested in learning where these formulas came from and why. She missed the beauty of mathematics.

In her second year at Carnegie Tech, Lenore changed her major from architecture to mathematics. She knew immediately that she had made the right choice.

Marriage and MIT

Lenore was now eighteen years old and she and Manuel married. They moved to Boston and rented a small apartment. The young couple's home was open to a growing circle of friends.

Manuel worked in Warren McCulloch's Neuro-Physiology Lab at MIT. Electrical engineers, mathematicians, philosophers, biologists, psychologists from all over the world came together there to work on a common problem—the understanding of the brain.

Lenore's and Manuel's apartment became a gathering place for the young people in the evenings. Eating, singing and playing drums in the small living room gave way to lively debates. One might recite a jazzed-up version of "Jabberwocky," a poem by Lewis Carroll. A poem by William Blake could start a discussion about infinity and other mysterious ideas that would last through the night.

Continuing College

Since she was now living in Boston, Lenore applied and was accepted at Simmons, a women's college. She hadn't the courage to reapply to MIT.

Mathematics classes at Simmons were not sophisticated enough. The college administrators agreed to send Lenore to MIT to take a course in modern algebra. The teacher was Isadore Singer.

"Here was a class with substance, depth, pace—everything I'd imagined a good course to be. It was hard, it was deep, it was abstract . I trusted this guy who was teaching it. He was a top mathematical researcher. The topics were important; they were leading somewhere. He wasn't just reading it out of a book." Lenore was transformed!

She was very quiet in class. She didn't know where she stood. She was just grateful to be there. The class was huge, nearly 100 students. Lenore finished the semester as one of the top students. Yet, when she applied to the graduate program in the Mathematics Department, she was told by the admissions officer, "MIT is no place for women. Here is a list of fine graduate schools. Apply to these," the man told her. "I would give my own daughters the same advice," he said.

Lenore was devastated. It looked as though she would be turned down again by the only place she wanted to go.

That weekend, there was a party at MIT and Professor Singer was there. He overheard a group of people discussing "the girl who wants to enroll in the graduate mathematics program."

"Who is she?" Singer asked. It was Lenore Blum. Professor Singer spoke up declaring she was one of his best students.

Within a few days Lenore received a letter of acceptance from MIT. Today, MIT brochures state, "MIT is a place for women."

Graduate School

Lenore decided she would start off graduate school with a bang. Most people took two or three courses each semester. Lenore would take eight. It was said that if a woman married she would not finish the program. Lenore was married and she would finish. It was said that a woman with a baby would drop out. Lenore would have a baby and remain.

At first, life was difficult. Eight courses were more than anyone could handle. Lenore dropped them one by one.

Manuel had his group, but she was on her own. Her co-workers were not used to working with a woman. Some even tried to drive a wedge between Lenore and Manuel.

Nevertheless, later on in the program, an older woman student spent a lot of time helping Lenore prepare for the crucial oral examinations. Lenore began to realize how women could help women.

A Fantastic Baby

She and Manuel had planned the arrival of their baby to coincide with the end of the school year. They took natural childbirth classes, which were quite unusual at the time, and on the last day of classes, went directly from MIT to the hospital to have the baby.

Lenore nursed her new baby, whose name was Avrim. He was " . . . so fantastic. We really had a lot of love for him right from the start." She was very happy.

Lenore, Manuel and Baby Avrim were always on the go. Child care centers were very rare. Babysitters were hard to come by. Fortunately, Manuel had an office in the basement at MIT and they cared for the baby there.

Working on a Thesis: Logic and Algebra _____

Lenore went on with her work. She did not wait for an advisor to set her to work on a topic. She found her own topic which became her thesis.

Some mathematicians were successfully using new methods of logic to solve old problems in algebra. What an intriguing idea! Lenore taught herself logic and carefully studied these methods and how they were applied. She wanted to understand why they worked. She thought and thought about this from many different angles.

But what to do with Avrim? He was now a toddler and needed a lot of attention. Fortunately, during this period, Lenore's mother was able to come to Boston to help care for him.

MIT had set aside a group of rooms especially for women students. This is where Lenore would work. The rooms were located right under the famous MIT dome in the main building. Here she made herself at home. She brought her papers and books and set up her working space.

Lenore became engrossed in study and thought. Sometimes she worked through the night. She bought food from the vending machines in the basement and slept in the lounge. Day turned into night. She lost track of time.

Slowly patterns started to emerge. Lenore began to see common features in the problems she was studying. She realized that one simple but powerful rule could solve them all. She understood this rule so well that she could explain it to the logic group in the mathematics department and show them why it worked.

This rule and its proof became her first theorem. Later she used this rule to discover new results in algebra herself. This work was to become her thesis, but first she needed an advisor.

A logic professor, Gerald Sacks, had recently come to MIT and was enthusiastic about Lenore's work. She became part of the logic group and was included in all their discussions and talks. Now she really belonged!

People who were respected in her field came to hear her work when she defended her thesis. The work she did earned her the doctoral degree. She received a post-doctoral fellowship and could go to work any place she chose for one year.

Berkeley was the obvious place to go. The mathematics department was one of the best in the country. A famous logician, Julia Robinson, lived there, and Manuel had a job offer there as well.

Berkeley, California

Politics was part of the Berkeley scene. It was 1968. People marched in the streets. They demonstrated against the war in Vietnam. They gathered to protest turning a park into a parking lot. They talked and they organized. Out of the Free Speech Movement came a new spirit. It was an exciting time.

Bright blue skies and Spanish tile roofs nestling in green hills reminded her of Caracas. People wore colorful, free-flowing clothing. Lenore felt she was coming home.

During her post-doctoral year at Berkeley, Lenore received several job offers. People at Yale were eager to work with her and offered her a position as assistant professor. She had offers from MIT and from Berkeley. Manuel had a job at Berkeley and so Lenore took the Berkeley offer, even though it was the lowest rank. The position was lecturer in mathematics.

Although discussions with her employers raised her hopes, no tenure or job security went with the position. What was worse, she had no professional group at Berkeley to support her career growth. These were important things to consider but Lenore did not realize this at the time.

About 1970. Lenore, Manuel and Avrim at Berkeley.

A Turn of Events

After two years as lecturer at Berkeley, Lenore was told she would not be re-hired. Lenore's talent and training prepared her to work at the highest level. But there were no women in positions at that level at any top mathematics department in the country. What to do?

About this time three concerned professors in the Berkeley Mathematics Department (Moe Hirsch, John Rhodes, and Steve Smale) sponsored a series of talks on mathematics and social responsibility. Lenore was asked to organize a panel on women and mathematics. She gathered together scholars Ravenna Helson, Sheila Johannsen, and Elizabeth Scott, who spoke about the history of women in mathematics and their present status. Several hundred people packed the lecture hall; this was the first such panel anywhere in the country. All of a sudden, Lenore became known as the expert on women and mathematics on the West Coast.

On the East Coast, women mathematicians had also begun to organize. That winter Mary Gray, of American University, led a protest at the mathematics meetings in Atlantic City. She wanted women to be part of the decision-making groups in the mathematics societies. During the spring, Mary issued a Newsletter and called for support. The Association for Women in Mathematics (AWM) was formed.

At first, Lenore was reluctant to join. She wanted to be known as a mathematician, not as a woman mathematician. But soon she became convinced: the situation for women in mathematics would not change without the AWM. Mary Gray was the first president of the AWM and Alice Schafer of Wellesley College the second. Later, Lenore became president.

The AWM and Change

It was 1971 and the beginning of a new era for Lenore and other women mathematicians. They spoke out. They wrote letters. They sponsored talks and panels and debates. In their newsletter, they pointed out how it was harder, sometimes impossible, for women mathematicians to get good jobs. They asked tough questions and they were not always popular. But they gained courage and support from one another.

For the first time, Lenore made friends with women who understood and valued her work. Two of these women, Judy Roitman, a logician and Bonnie Miller, an astrophysicist, were particularly important for Lenore.

"Without their friendship, I could not have accomplished what I have," she says. "When I felt down and out, they boosted my spirits. And when things went well, they cheered me on. Even today, we keep in close touch, though we now live in different parts of the country."

The situation began to change. By the middle of the 1970's, women mathematicians were becoming more visible. They were invited to present their research results at important professional meetings. They were elected to high positions in the professional associations. This kind of activity is very important for the career growth of women mathematicians.

By the end of the 1970's, women mathematicians were getting better jobs, some in top departments. Indeed, by its tenth birthday, the AWM had brought about many important changes for women in mathematics.

A New Direction

In the fall of 1973, Lenore was hired to teach a class in college algebra at Mills College, a women's college in Oakland, California. She thought the course was dull. It repeated high school work and didn't seem to lead anywhere. In the middle of it she said, "I am going to teach you something much more useful—calculus!" Then and there, she realized that a carefully designed "pre-calculus" course could open doors for many women.

Some girls drop mathematics in high school because they think it's hard and boring. They think it's not important for their future. Then, when they get to college, they want to enter fields like medicine, business, engineering and computer science, which require a knowledge of higher mathematics. A good pre-calculus course can introduce them to calculus and higher mathematics. It helps bridge a gap for women students.

Lenore helped her students feel the excitement of mathematics and helped them gain confidence so they could really do it. She taught them to look for pictures in equations. Is it a straight line? Is it a rising curve? She showed them that people solve problems in different ways. Sometimes, guessing and checking your answer is a good way.

She said, "You can learn these things; you don't have to be born a math whiz."

Lenore always knew that what she was saying was much bigger than this one class. This one class grew into a program. Lenore became Head of the new Department of Mathematics and Computer Science at Mills College.

Students at Mills started to study calculus, computer science, statistics—courses that would help women enter fields that they did not even think of in the past. Some new fields for women to consider were science, engineering, medicine and economics. By 1980, more students at Mills were taking classes in mathematics and computer science than in any other subject. They were also getting jobs in these technical fields.

The Math-Science Network

Meanwhile, progress was being made on other fronts. At the Lawrence Hall of Science in Berkeley, after-school classes in mathematics and science were being offered for grade school students. Nancy Kreinberg, a director of the program, noticed that very few girls were taking these classes. She decided to do something about this. She and some colleagues started Math for Girls.

Math for Girls was a special class for six to twelve-year-olds who got together to solve puzzles and play challenging games. These activities were designed to develop important problem-solving skills. Their teachers were college women who enjoyed mathematics, science and engineering.

Nancy worked with young girls and Lenore worked with college women, but soon they

found out both groups had a lot in common. One summer afternoon in 1975, they met with like-minded scientists and educa-tors to share ideas. Quickly, this group realized they were onto something big. They could combine many resources and talents in the community and create a powerful force for change. The Math/Science Network was on its way!

Lenore and Nancy took their message across the country.

Lenore in her Berkeley office, early 1980's.

"Encourage young women to take as much mathematics as they can," they said. "It's important for their future." Thousands and thousands of young women heard this message as they attended Network events such as the *Expanding Your Horizons* conferences.

It was thrilling to make important things happen. But where was the little girl who liked to go off, do mathematics, make drawings, and dream?

Pulling in the Reins _____

With the coming of the 1980's, Lenore pulled in the reins. She decided to return to her work as a research mathematician. This decision was the beginning of a new exciting journey.

Taking a leave of absence for a year from Mills College, Lenore returned to MIT. Manuel and Avrim, her family, stayed on the West Coast, where Lenore visited them frequently.

Lenore Blum working with students at Mills College.

Back to Research: Mathematics and Computer Science

In her research, Lenore uses mathematics to study why some problems are hard for computers and why some are easy—and why some problems can't even be solved at all!

Several years earlier, she and Manuel had written a paper together. They were interested in designing computers that could learn from examples in much the same way young children do. Lenore's new work explores these ideas further.

Research goes very slowly. "Some days," Lenore says, "I may sit at my desk for hours and scribble only a few lines. But each day, my understanding increases, and I begin to see things fall into place. I view my work as the beginning of a ten-year program."

Finding interconnections between her work and other projects, working among creative people who value her work—these are Lenore's rewards.

"It's exciting," she says, "to talk with other people and see how what we do fits together."

Flourishing New Directions

Over the years her research career has flourished. The invitation to talk about her work to the entire mathematics world at the 1990 International Congress of Mathematicians in Kyoto, Japan meant her work was recognized as important.

Although developing her new theories has been the cornerstone of her work in the eighties, Lenore has been involved in many other important things. She became vice-president of the American Mathematics Society. She has been involved in setting up an electronic communication link with African mathematicians. In 1991 Lenore presented an exciting history talk at The Association for Women in Mathematics' twentieth anniversary celebration.

In the fall of 1992 Lenore began a new career as Deputy Director of the Mathematical Sciences Research Institute in Berkeley, one of the foremost mathematics research institutes in the world. One of the exciting things about this new job is the opportunity to work with Nancy Kreinberg again. The Institute is situated on a hill just above the Lawrence Hall of Science where Nancy

Lenore with other women mathematicians at Pan-African Congress of Mathematicians, Nairobi, summer of 1991

Young women from the Mills Summer Mathematics Institute share their experiences at the 1992 Baltimore meeting of the Association of Women in Mathematics (AWM). The Mills program is a summer mathematics program for women co-directed by Lenore Blum and supported by the National Science Foundation. From left to right: Nancy Cunningham, Rebecca Field, Cheryl Grood, Jessica Wolpas, Min Kang, Maria Basterra, Sunita Vatuk, Professor Lenore Blum, Kendra Hershey, and Julia Kerr.

works. The two women look forward to exploring new ways to link the worlds of mathematics educators and mathematicians in order to continue to encourage the participation of women in the world of mathematics.

The Math-Science Network Today

Lenore helped start the Math/Science Network, a San Francisco-based group. The Network encourages young women to study mathematics in order to qualify for many careers, such as engineering, medicine, and computer science.

The Math/Science Network meets once a month. At the meetings people talk while seated around a horseshoe-shaped table. Let's pretend that there are ten people at this meeting. Who is sitting where? Can you figure it out?

Lenore is seated to the left of Len (not necessarily next to). Elizabeth is to the left of Lenore. Nancy is between Rita and Len, with Rita on her right. Elizabeth is to the right of Diane and Carol is left of Len. Carol is between Jan and Elizabeth. Flora is between Lenore and Elizabeth. Oh yes! And Lucy comes in late, walks to the end of the table, and sits down next to Rita.

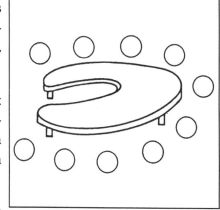

Move names on slips of paper. It's much easier that way!

Activities

• Shade all multiples of 3

Hint — All multiple of 3 have a digital sum that is 3, or 6, or 9.

More on digital sums

• 26 is a two digit number. Its digits are 2 and 6. Its digital sum is 8 since 2 + 6 = 8.

• 45 is another two digit number. Its digits are 4 and 5. Its digital sum is 9, since 4 + 5 = 9. (45 is also a multiple of 3 since its digital sum is 9.)

• To find the digital sum of 458, add 4 + 5 + 8 to get 17. Since 17 is not a one digit number it is not the digital sum we need. Continue to add 1 + 7 to get 8 which is the digital sum of 458. (Remember, the digital sum we want to look at is a one digit number from 1 through 9.)

• To find the digital sum of 525 add 5 + 2 + 5 to get 12. Then add 1 + 2 to get 3 which is the digital sum of 525. Since 3 is a multiple of 3, 525 is also a multiple of 3. (If you don't believe this, divide 525 by 3 to convince yourself.)

Practice your new, fast, test for multiples of 3.
Shade all multiples of 3 to complete the picture of Caracas below.

Drawing by by Analee Nunan.

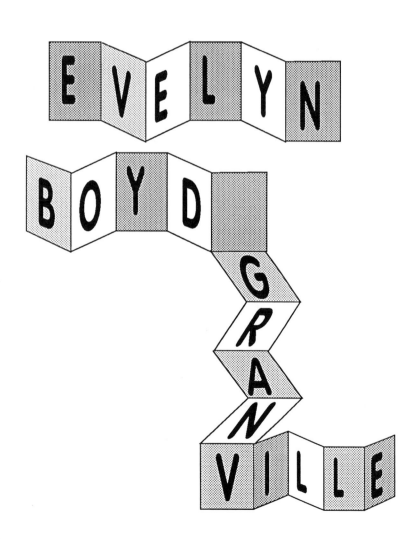

EVELYN BOYD GRANVILLE
1924-

"The library was such a pleasant place...

so quite and serene..."

An Unusal Journey

When Evelyn Boyd was eighteen years old, she headed north to Smith College in Northampton, Massachusetts. This was an exciting trip for the young woman, for she had rarely traveled any distance from home before. She was not worried though. After all, she was traveling with her mother, Julia, to whom she was very close, and her Aunt Louise.

Louise was her mother's twin sister. Louise had never married and was almost a second mother to Evelyn. They traveled in Aunt Louise's brand new 1941 Dodge, so new, in fact, that Aunt Louise was not yet comfortable driving long distances. The women were being chauffeured by a cousin who had come all the way from Detroit to Washington, D.C., to drive them to Massachusetts. Evelyn grew up in this kind of family. It was a family where people helped each other. In fact, they went out of their way to help each other.

What was Evelyn thinking about as she was driving to college? Perhaps she was thinking about the comfortable world she was leaving behind—her home, her friends, school and teachers who had been so supportive and inspiring.

Evelyn and her family were of very modest means and Black. She grew up in an all Black neighborhood. Although this neighborhood was just a few blocks from some of the most elegant parts of Washington, D.C., Evelyn felt neither envy nor that she was different. There was a lot of caring in her family and the world seemed like a rich place.

When she was very little, she lived in a tiny apartment in a building where her father was a caretaker. Later, when he separated from her mother, her mother supported the family on a modest, though secure, salary. Julia was a civil service worker in the U.S.

Government Bureau of Engraving and Printing. Although her father continued to visit the family from time to time, it was her mother who supported them.

As a child, Evelyn was shy. She looked up to her sister Doris. Doris was less than two years older than Evelyn, but to the little girl with the big brown eyes and black braids, Doris was big and important.

Summer Fun

Summers were an especially happy time. There were many things to do at the neighborhood playground. Evelyn joined the arts and crafts program and played games. But what she liked most of all was the library.

"The library was such a pleasant place; so quiet and serene. I was fascinated by China, and the Chinese people, and my favorite book was *Yung Fu of the Upper Yangtse*. I read it over and over."

Evelyn and Doris often spent a month in the country with Miss Lucy Booker, an old family friend. Miss Lucy Booker cleaned Washington houses in the winter and spent the summers at a farm she and her husband owned in Linden, Virginia, in the country. Sometimes she boarded children, such as Evelyn and Doris, from the city. Life on the farm was a pleasant change. Evelyn and Doris played with the dogs and pigs and helped care for them. Since there was no electricity on the farm, Miss Lucy put the milk and butter in the creek to keep it cold.

First in the Class at Dunbar High

Back in Washington, Evelyn attended neighborhood schools. The time was the 1930's and schools were segregated. All the teachers and students in Evelyn's school were Black. School was a happy place for Evelyn. She was bright and she didn't mind working hard. In mathematics, she was at the top of her class and she did well in other subjects too. When she graduated from junior high school, she was salutatorian of her class, which means second in class standing. At the end of senior high, she was valedictorian—first in her class!

The students at her high school, Dunbar, were hard-working. Success meant a lot to them. Most of them went on to college, and many of these colleges were the Ivy League colleges such as Smith, Vassar, Mt. Holyoke, Amherst, and Wellesley. The teachers at Dunbar were strict and had high goals for their students. They inspired the students to achieve at the highest levels. Evelyn's math teacher, Miss Mary Cromwell, suggested Evelyn apply to Smith College. Two relatives of Miss Cromwell had graduated from Smith.

Evelyn applied and was admitted, but she was given no financial aid the first year.

Evelyn's family had little money, but they had faith in themselves. Her mother had always saved as much as she could for her girls' education. She contributed $500 toward Evelyn's freshman year expenses. Aunt Louise contributed another $500 to help her favorite niece. There was another $100, which was one-fourth of a scholarship from Phi Delta Kappa, a national sorority for Black women. Altogether, Evelyn went off to Smith with $1100 for that year's tuition and living expenses.

Life at Smith College

Classes were no problem for her. She wanted to major in mathematics and physics, and she did. Her first encounter with astronomy gave her a passion for the subject. In fact, it was her astronomy professor, Miss Williams, who helped Evelyn get her first summer job in industry, after her freshman year.

Socially, Evelyn felt she was accepted there. Smith was a college for women so Evelyn, who had never dated boys during high school, did not feel left out of the social whirl. It was World War II, many young men had gone to war, and there was little weekend partying at nearby men's colleges.

As far as expensive clothes were concerned, the young women at Smith wore jeans to class and on weekends. After her first year, Evelyn lived at a co-op house, where students took turns sharing the chores. Living there was less expensive than at the regular dormitories.

When Evelyn was still in high school, she planned to become a teacher. This profession was highly respected within the Black community. Now that she was at college, though, Evelyn started thinking about other possibilities — perhaps becoming a physicist or mathematician in industry.

After Evelyn graduated from Smith in 1945, it felt quite natural to go on to earn an advanced degree in mathematics. She received a fellowship from Smith College, and financial aid from Yale, and with these she was admitted to Yale University to study for a doctoral degree. She received Yale fellowships for the next two years and during the last year was an Atomic Energy Commission postdoctoral fellow.

Setting a New Record

It is not difficult to imagine what a triumph this was for the young woman. She herself did not seem to grasp how outstanding an accomplishment it was. It was astounding that a woman, a Black woman, be admitted to such a program at that time in the history of our country. Although she was not aware of it at the time, Evelyn, together with Marjorie Brown at the University of Michigan, were the first Black women to receive doctoral degrees in mathematics. The year was 1949. Evelyn set a record of which she was not aware until much later in her life.

Yale was lots of fun. Evelyn had a circle of friends, young men and women, and enough money for occasional train trips to New York for a day of shopping and theater. She studied hard, but also had time to enjoy campus life.

Evelyn graduated from Yale with her Ph.D. in mathematics and a college physics major behind her. She was not personally aware of discrimination against Blacks and women as she looked for her first job. The United States was just launching its space program and Evelyn's background was tailor-made for jobs in this program. Still keeping her mind open to possibilities in college teaching, Evelyn had an interview at a New York city college. Years later, she was told by a colleague the "behind the scenes" truth about that interview: because she was a Black woman, the interviewers had never seriously considered her for the job, even with a Yale doctoral degree.

New York—Not Like Home

Her first job was at New York University's Institute of Mathematics and Science and of course Evelyn had to move to New York. To her, the city was all noise, dirt and concrete. Apartment hunting brought her face to face with "de facto" housing discrimination. This means that, although housing and apartments were not segregated by law, old attitudes on the part of landlords caused Evelyn to have a very hard time finding an apartment to rent. After looking and looking, she moved in with another Black woman who was a friend of her mother's.

Settling Down at Fisk

After a year, she decided to leave New York City for a job she really wanted, teaching at Fisk University, a well-known college for Black women and men in Nashville, Tennessee. Evelyn went to work at Fisk as a mathematics instructor.

At the opening of the school year, the President of Fisk introduced the new faculty. When he came to Evelyn and described her education and achievements, there was a gasp from the audience.

Speaking of herself, Evelyn says, "It was wonderful for a woman to have that background. That in itself was an inspiration to the students."

Evelyn taught undergraduate mathematics courses for the next two years. Several women students later mentioned Evelyn in their doctoral dissertations, giving her credit for inspiring them.

Evelyn enjoyed being with people. She was discovering this over and over. At Fisk, she lived in a large house with other members of the faculty. She liked the close interaction with her co-workers.

The Space Program—Fascinating Work

In the 1950's Evelyn worked at IBM and was closely involved with Vanguard, the first missile program in the United States. She worked at the Vanguard Computer Center, sometimes in New York and sometimes back home in Washington. The work was fascinating and made good use of her training in functional analysis. Answers describing the inside workings of the rockets were fed into computers to guide the Vanguard satellite, and Evelyn made "on-site" calculations at the main viewing area when the satellite was first launched. This was the most interesting job she had in industry.

Evelyn Boyd Granville at IBM Vangard Computing Center, Washington, D.C. circa 1958.

One year, on a summer vacation, Evelyn's life took a sudden turn. She had gone to southern California to visit a friend and there she met and later married her first husband. He was a minister in the Community Church. In 1960, Evelyn, now married, moved from the East Coast to Los Angeles. She was 36 years old.

Using her long experience in the space program, she applied and got a job at Space Technology Labs (STL) as a mathematical analyst, one who studies rocket trajectories. She was one of few women and the only Black at this job. As before, she was well treated by her male co-workers. Still, they had their own social group and she had none.

Making Friends and Influencing People

How was she to make friends in the brand new city? One way was through the church. As a young girl and a teenager, she had always been very active in her church. Now she joined her husband's church. He was part of a group of ministers who raised funds for Martin Luther King Jr., the celebrated Black leader of the Civil Rights Movement. Evelyn's husband helped to organize large rallies when Dr. King came to town. Evelyn met Dr. King several times and thought he was a fascinating person.

During this period, Evelyn was asked to speak three or four times a year at Women's Day Programs at churches. These programs were held in connection with civil rights events. Though she was a little nervous as a public speaker, she learned to carry off her speeches gracefully.

By the end of the 60's, Evelyn was again working at IBM. People all around her were being transferred, because IBM had not won a large contract from the government for work in the Los Angeles area. Evelyn's supervisor, who did not want the company to lose her, urged her to transfer to IBM in northern California or back in Washington. But Evelyn didn't want to go.

The Right Time and the Right Place

By this time, she was divorced. She lived in the View Park area of Los Angeles and owned her own home. Perhaps, she thought, this was the time to think about changing careers. She would leave industry and re-enter the world of college teaching. She had always loved teaching.

Evelyn applied to California State University at Los Angeles, close to home. Just as her job at IBM drew to a close, she heard she had gotten the job at Cal State University at Los Angeles.

"I have always been in the right place at the right time," she says.

At Cal State, Evelyn taught mathematics for teachers and courses in computer science. Teaching had its good points and its bad ones. Students at Cal State seemed far less eager to learn, than those she remembered from Dunbar High School and Fisk University. These students were weak in math and teaching them was difficult.

On the bright side, she was practicing the profession she loved."I have always loved to explain things, and see someone learn from that," she says.

Although she earned exactly half her old salary in industry, she joked that the only day she missed her old job was on payday.

About this time, she began a very rewarding project. She and a co-worker at Cal State, Jason Frand, decided to write a book. The two took five years to complete it, testing it on their own classes, revising it as they went along. She remembers many grueling evenings and weekends writing and rewriting and rewriting again. The title of their book was *Theory and Applications of Mathematics for Teachers*. Although now out of print, it was well received, and the publication went into a second edition.

Evelyn became a full Professor at the university. She had published a successful book in her field, and had taught for many years. She would have liked, in addition, to enter college administration. During the 1970's, under pressure from affirmative action laws, the status of women, especially minority women, rose in both industry and higher education. At Cal State, minority women were offered a one-year training program to allow them to become deans, department heads and administrators. In other words, to be in positions where they could influence policy.

"If I were younger," Evelyn says, "I would enter this program."

Marriage

In 1970, Evelyn married Ed Granville. Ed worked in real estate and banking. They were very active in their church. Evelyn kept the church financial records. She loves to sing and was a member of the choir. The Granvilles attended Bible Study every week and had many friends among church members.

Moving to the Country

In 1984, Evelyn Granville retired her professorship and moved to Tyler, Texas. Together with Ed's relatives the Granvilles bought a farm, and built a house. After life in Los Angeles, with its crowds and fast pace, Evelyn enjoyed the change to a more peaceful, easygoing lifestyle.

Evelyn has always had a spirit of adventure and a strong curiosity. And she herself has surely learned how to "stay on her toes". It is these qualities that brought to life for her the story of Yung Fu, the Chinese boy. It is this that marked her attraction to astronomy with its study of outer reaches of the universe, that prompted her job changes, her moves and her change of careers.

Is the Texas move her final adventure? The last we heard she was still teaching mathematics and computer-science at a Texas college. And she still was finding time to raise chickens. Final change or not, the Texas move certainly brought Evelyn Granville close to nature, family, teaching and the community — all the things she cares the most about.

Activities

Evelyn Granville enjoys story puzzles such as this one:

> **A bus load of students are on a trip to a school at the other end of town where they are presenting an evening of entertainment.**
>
> - **Fifteen students in the band are going to perform.**
> - **Eight students are in a play.**
> - **Three students who play in the band are in the play as well.**
> - **Ten students are in the chorus.**
> - **Of these, two are also in the play.**
> - **No students are in both the band and the chorus.**
> - **Six students are just going along for the ride.**
> - **Two teachers are going along to maintain law and order.**

?

How many seats are needed on the bus, not counting the driver?

A One way to solve this problem is to fill in the diagram below. Show the number of students in each activity—band, play, chorus. Show the number of students in more than one activity where the circles overlap. Then continue with the diagram and labels as you have them.

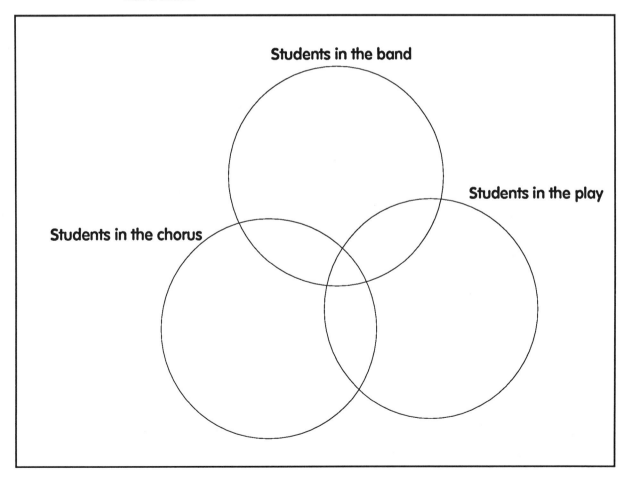

Students in the band

Students in the play

Students in the chorus

The Pascal Triangle

One of Evelyn Granville's favorite mathematical patterns is the Pascal triangle.

The diagram below shows a partly completed Pascal triangle. Notice each number is the sum of its two closest neighbors in the row above.

Look at the triangle and see if you can continue the pattern for at least two more rows. (Actually the Pascal triangle can be made as large as you like.) It may help to notice that the Pascal triangle is symmetric around line AB. That means if the Pascal triangle is folded along a vertical line down its middle (line AB), the numbers on both sides will be the same.

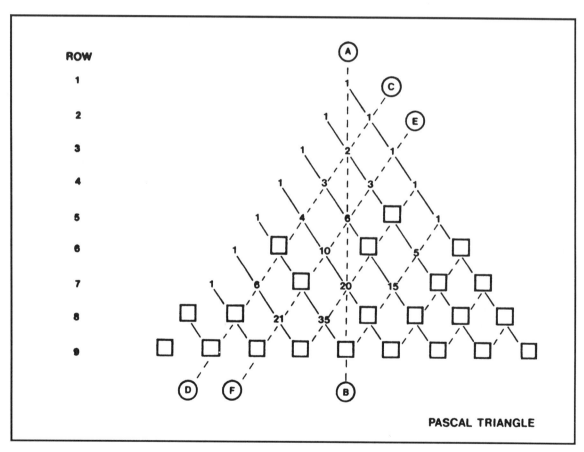

PASCAL TRIANGLE

The numbers that make up the Pascal triangle turn out to be a good description of activities like coin tossing. For example: **If you toss one coin, only two things can happen; you can turn up one head or one tail.** These are the ones in Row 2 of the Pascal triangle in the diagram.

experiment —
Toss two coins and four things can happen.

result —

HEAD	HEAD	1	[2 HEADS}
HEAD	TAIL	2	[HEAD-TAIL] combinations
TAIL	HEAD		
TAIL	TAIL	1	[2 TAILS]

This can be summarized as

1 [Both HEAD]   2 [HEAD,TAIL] combination   1 [Both TAIL]

Notice these totals (1,2,1) in Row 3 of the Pascal triangle.

experiment —
If you toss 3 coins, 8 different things can happen; you can get

result —

HEAD	HEAD	HEAD	1	[3 HEADS]
HEAD	HEAD	TAIL		
HEAD	TAIL	HEAD	3	[2 HEAD, 1 TAIL] combinations
TAIL	HEAD	HEAD		
HEAD	TAIL	TAIL		
TAIL	HEAD	TAIL	3	[1 HEAD, 2 TAIL] combinations
TAIL	TAIL	HEAD		
TAIL	TAIL	TAIL	1	[3 TAILS]

When this is summarized, you get the totals in Row 4 of the Pascal triangle.

That is: 1 [ALL HEAD]   3 [2 HEAD, 1 TAIL]   3 [1 HEAD, 2 TAIL]   1 [3 TAIL]

For 4 coins, 16 different things or events can happen. They would look like this:

result—

```
H   H   H   H   1 [4 HEADS]

H   H   H   T
H   H   T   H
H   T   H   H
T   H   H   H   4 [3 HEAD, 1 TAIL] combinations

H   H   T   T
H   T   H   T
T   H   T   H
T   H   H   T
H   T   T   H
T   T   H   H   6 [2 HEAD, 2 TAIL] combinations

T   T   T   H
T   T   H   T
T   H   T   T
H   T   T   T   4 [3 TAIL, 1 HEAD] combinations

T   T   T   T   1 [4 TAIL]
```

Complete the number for each situation described below.

> _____ 4 HEADS
>
> _____ 3 HEADS, 1 TAIL
>
> _____ 2 HEADS, 2 TAILS
>
> _____ 1 HEAD, 3 TAILS
>
> _____ 4 TAILS

How many different events happen here?
What row of the Pascal triangle shows this? (See Pascal diagram)
How about 5 coins? How many different events can happen? Can you enumerate all the events when 5 coins are tossed? What row of the Pascal triangle describes this?

If you think about it you will notice many other interesting patterns in the Pascal triangle. For example: Notice the 1's along the outside edge on either side of the triangle.

Now look at the numbers in line CD. Notice these numbers are the sequence of counting numbers (1, 2, 3, . . .).

Look at the numbers in EF in the Pascal triangle. These are called the triangular numbers. Look at the diagram below and you will see why.

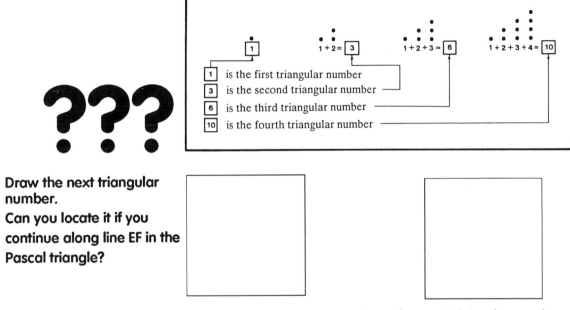

Draw the next triangular number.
Can you locate it if you continue along line EF in the Pascal triangle?

Draw the next triangular number.
Can you find it on line EF as well?

Here is another pattern in the Pascal triangle. Add the numbers across each row in the triangle and fill in column B. Then complete column C. Notice the pattern of numbers in column B.

Each number is a power of 2. Complete column D.

This is just a sampling of the kinds of patterns that have made the Pascal triangle such an ongoing source of interest to people who enjoy mathematics.

A row	B sum across rows	C multiples of 2s	D powers of 2
1	1	1	2^0
2	2	2=2	2^1
3	4	2x2=4	2^2
4	8	2x2x2=8	2^3
5	—	___	___
6	—	___	___
7	—	___	___
8	—	___	___
9	—	___	___
10	—	___	___

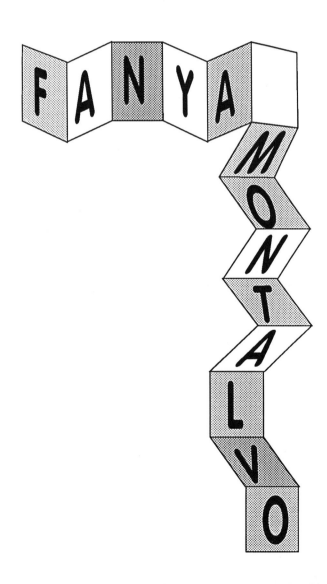

FANYA MONTALVO
1948-

"Don't go too fast but I go pretty far,
For one who don't drive I've been all around the world,
Some people say I've done all right for a girl..."

Fanya watched her pet duck, Oscar, waddle across the narrow concrete yard. She had just returned from walking Oscar around the neighborhood where kids had stared at the big white duck. Fanya loved the sensation she and Oscar created. She was building a corral for the duck, and she picked up her hammer and nailed a strip of wood.

Fanya was eleven years old. She liked to build things. If her family couldn't afford to buy something, she built it out of whatever materials were around. With cardboard boxes, pegboard, and paint she had made herself a desk. For her friend and herself she had made a dollhouse out of cardboard.

Dolls, Ducks, and a Difference _____

Dolls didn't really interest Fanya. She was different from some girls that way. Different in other ways too. Oh yes, her apartment house looked like all the others on the street, and that street looked like all the others in Logan Square, her section of Chicago. Her duck, she supposed, looked like any other white duck pretty much. It wasn't any of those things. It was she herself. Her skin was more olive than most, and her language was not quite like the others in her class. Fanya understood textbook English, but it was hard to understand expressions, slang and jokes. She often felt left out if she didn't understand what was being said. Dollhouses and corrals were easy, but not the English language.

Five years before, when Fanya was six years old, she had moved with her family from Monterrey, Mexico. Fanya's father was a radio and TV repairman. When Fanya was only one, he had come to the United States to find work. Staying there, he worked hard and saved his money. With a friend as his partner, he bought an apartment house in Logan Square. This section of Chicago had no Mexicans. Here he moved his wife, three sons and his daughter, Fanya.

The family returned to Monterrey for vacations. The trip in the family Dodge took about three days. Monterrey was due south of Chicago about 1300 miles.

In the United States, the Montalvos spoke Spanish at home. On this their father insisted. He wanted his children to be proud of their heritage. When they went to Monterrey, they renewed contact with that heritage. They had fun with their relatives and celebrated.

In the United States, the family kept to themselves. Still, Fanya didn't want to move back to Mexico. Her life was in the United States now.

In school she had to learn English on her own. When she went to first grade, the teacher just talked louder when Fanya didn't understand her. Reading was very hard, and it took Fanya a long time to finish a book. She really had to struggle with this new language.

Numbers, though, were the same in Spanish and English, and mathematics was much easier for her. Problem-solving was fun. The best fun, though, didn't happen in school at all. The best fun was when her father made up puzzles for the four children to solve at the dinner table. Fanya was six years younger than her next older brother, yet sometimes she got the answer first. When that happened, they would look at her and say, "That little one; she is clever!" Her mother was proud of her daughter who was so good at reasoning.

The Physics of Things

Fun was also jokes with her brothers—teasing and playing with them. Her brother Ramiro began to study physics in college. He asked her questions about the physics of things. He told her about the theory of relativity.

"Suppose you had a spaceship going at three-quarters the speed of light and another ship goes by it at a passing speed of three-quarters of the speed of light," he would say. "How fast is the second ship going?"

"One and a half times the speed of light," Fanya would answer.

"No!" Ramiro would say, "Nothing can go faster than the speed of light."

Then he would try to explain to Fanya why that was so. She was very excited by it.

About this time she read a book called *One, Two, Three, Infinity*, by George Gamow. It was a book about physics for teenagers that had "wonderful puzzles and paradoxes in it."

Mathematics was fun too. Without trying very hard, Fanya got 100's on quizzes. She liked to work things out for herself with no help from others.

A Kitchen Experiment

When she was in seventh grade, Fanya showed the class at school an experiment she had worked out at home. She had made a rocket from aluminum foil and a wooden matchhead.

This was the way she made it. She took the head off a flat kitchen timer and inserted a butter knife in the slot of a shaft so that when she set the timer, the knife rotated slowly. She placed a lighted candle so that at zero, the knife would move over the candle and become heated. At the end of the knife handle she constructed a small platform with a bobby pin, and set a tiny "rocket" on the platform. The rocket was a wooden matchhead with the stick broken off, and an exhaust channel made with a pin in its base. The matchhead with a pin stuck in it was wrapped tightly in foil. The exhaust channel was formed by the foil surrounding the pin. When the pin was removed, the knife moved over the candle, heated up, heated the space inside the "rocket" and— blast off! It flew up and halfway across the classroom.

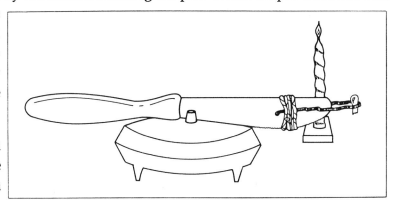

"It was great," Fanya recalls. It wouldn't flame very much. Just go with a soft `whoosh'.

"Some of the kids in the class were surprised that I could do experiments with fire. When I became interested in candles and matches at a younger age, my mother let me play with them in front of her until she knew I could handle them safely. I was proud of my parents for this." (IMPORTANT! Parents or responsible adults must check all experiments with fire and explosives, before they are set off.)

Fanya attended a Catholic girls' high school. She made friends with a classmate named Sue. Sue and Fanya are friends to this day. "She is my oldest, dearest friend," says Fanya. Sue was fun. She liked boys and they liked her. Fanya and Sue hung around together. They went to each other's house after school. They went to the beach on Lake Michigan in the summer. They listened to records and went to dances. Sometimes they double-dated.

Sue taught Fanya how to use make-up. "I felt my looks were totally non-standard," Fanya recalls. "The standard look was Anglo—blond and beautiful. I was dark and swarthy and flat-chested and skinny. I was everything you weren't supposed to be in the late 1950's and early 60's. I didn't feel very attractive at all in those years."

Moving Up in Math and Science

Because of her slow reading, Fanya tested poorly in high school entrance tests. She was placed in the lowest academic track. But she immediately began to move up in mathematics and science. It took awhile longer in English. Fanya remained fluent in Spanish, her native language. In English she worked hard, and by her senior year moved to the middle track. In science and mathematics she was in the top group, taking honors math.

In geometry class, Sue asked Fanya for help and Fanya tutored her. Sue got an "A" on the next test. The teacher called Sue in and accused her of cheating. "No," Sue explained, "Fanya has been working with me."

Fanya felt very proud but she didn't really work hard herself. She got by with B+'s. It took her some years before she could discipline herself to work hard for a goal she set herself. As a student, she didn't take herself that seriously.

Fanya followed in Ramiro's footsteps at college and majored in physics. She attended Loyola University, a Catholic university in Chicago. There were no other girls in physics but, Fanya reasoned, "I'm already different. I'll go into this field."

Fanya loved physics. What are things made of? Why does heat rise and cold sink? What makes light travel? What rules govern the universe?

Problems of a Professional Woman

By her sophomore year, the problems of being a "professional technical woman" were beginning to surface. First of all, there was a huge change at home. Ramiro, ten years older than Fanya, left home that year. Gone was her supportive older brother. Gone her intellectual friend. Fanya felt abandoned.

She wanted to make friends with the "smart kids" at school, but she felt awkward around them. She was from a humble background. Was that why she felt shy around them?

In Fanya's physics courses there were only boys. These boys were not helpful the way her brothers had been. They weren't available to work on homework assignments. The professors really didn't act helpful either.

"Well," she thought, "the whole physics department is too cold." She decided to choose another major, something more humane. She switched her major to psychology.

With her strong mathematics background, Fanya went into mathematical psychology. After a psychologist chooses a question about human behavior to explore, a mathematical psychologist takes that problem and finds ways to measure it. Fanya, for instance, learned to chart and graph statistics in her courses.

Fanya still did not think she was anything special as a student. In fact, when a professor counseled her in her senior year to enter graduate school, she argued with him:

"Look. I want to graduate, get a job, get an apartment and have my own money. I don't know what branch of science I'm really interested in. Basically, I want to find a man who is a scientist and smarter than I am, marry him, and raise a family!"

The professor laughed at her and referred her to another psychology professor named Naomi Weisstein, who was teaching a course about sensory perception the coming semester. Fanya signed up for her course and Naomi Weisstein helped her change her life.

A New Way of Life

Fanya joined Naomi and her circle of friends. This group was high-spirited and liked to have fun as well as work hard.

Naomi recognized Fanya's talent and offered her a job. After graduation, Fanya became head of Naomi's vision lab at Loyola University. At last she had a job, her own money, and her own apartment.

During the two years she held this job, she had time to think about her goals. She needed the freedom to explore her love of science. Marriage could wait. She wanted a full chance to try out her new life before looking for a partner.

As head of the lab, Fanya helped design and set up experiments in vision. Human "subjects" would sit in darkened little rooms for hours and look through tachistoscopes. These machines flashed visual stimuli or pictures for one-fifth of a second or less. The object of the experiments was to learn how human beings see objects and translate what they see to the brain. Fanya and Naomi published a paper together about these experiments.

After two years, Fanya was getting bored. "Go to graduate school," Naomi urged.

Fanya still wasn't sure she was good enough to go on, but with support from Naomi, her new friends and co-workers, she enrolled in a masters program at the University of Illinois in Chicago.

During this year she studied all sorts of subjects that would be very useful later in her studies and work. The courses were in television, signal processing, systems theory and frequency analysis. The people in her classes were mostly electrical engineering students.

The CWLRB

About this time, Fanya, Naomi and their friends decided to start a women's rock band. Rock music was performed mostly by men, but the "Chicago Women's Liberation Rock Bank" or CWLRB would change all that!

Fanya wanted to sing, but as a band soloist she wasn't very good. "Do something else," the others said.

Someone had given up the drums and they were just sitting there. "I started to play them just to keep time for the rest of the band," she says. "Hey, you're pretty good!" they said. And that's how she became the drummer of CWLRB.

> *Don't go too fast but I go pretty far,*
> *For someone who don't drive I've been all around the world,*
> *Some people say I've done all right for a girl . . ."*
>
> —Lyrics from "I've Got a Brand New Pair of Rollerskates" by Melanie

Fanya loved performing. When she entered the University of Massachusetts (U-Mass) Computer Science Department to earn her doctorate degree, performing came in handy. Giving papers at conferences was a top requirement for the doctoral students, and Fanya could put on a good show.

U-Mass hadn't been Fanya's first choice, even though it was a very good school. The top schools in her field—MIT (Massachusetts Institute of Technology) and Stanford University—had turned her down. Her record was not quite good enough.

The head of the department at U-Mass was a friend of Naomi's named Michael Arbib. He accepted Fanya as his research assistant. She became one of a select group called "Arbib's Kids."

Out of eighteen doctoral students only two were women and the other woman did not arrive until Fanya was nearly finished with her graduate work. But taking courses in computer science at U-Mass was not like the physics courses back in college had been. Here, she got equal treatment with the men students. She was part of the group.

At U-Mass, Fanya had all the ingredients for success: a close group of co-workers, a leader who was inspiring and fair, and an atmosphere that was open to new ideas and new people. Fanya blossomed at U-Mass.

She no longer felt unattractive. Her olive skin, intense brown eyes, and dark hair were no longer unusual. Blond hair and blue eyes were not the only standard of beauty. Times had changed.

Dr. Fanya S. Montalvo and AI

At the end of four years Fanya earned her doctoral degree. Now she was Dr. Fanya S. Montalvo. She was ready for new adventures and wanted to live in a new place. Perhaps California.

When Fanya had worked in Naomi's vision lab, Naomi introduced Fanya to a branch of Computer Science called Artificial Intelligence, or AI as it was known. Researchers in AI study how the brain works, and try to design computer programs that act like people. Fanya was fascinated. She looked for a job in this field.

But to get a job in AI "you needed to come out of one of the top three universities with AI labs," she explains. Instead, she took a job in California at the Lawrence Laboratory at Berkeley.

Her position at the lab was a postdoctoral one. She continued the work she had done on vision in her doctoral thesis by applying this work to computer graphics.

She kept her eye on developments at the Stanford AI Lab. Stanford was fairly close to Berkeley. She sat in on a course at Stanford. Her teacher had constructed a computer language called FOL, which meant first-order logic. The language did not use numbers, nor did it use words. Instead, it used symbols such as those taught in the study of logic, a branch of philosophy. Her teacher built a structure in the computer that could move these symbols around.

"A lot of what he was thinking about was what I had been thinking about," says Fanya, "so I invited him to lunch to talk about it. Secretly, I was hoping he would offer me a position at Stanford AI Lab.

"Instead he said, 'They're looking for people to do this kind of work at Hewlett-Packard Laboratory.' "

Fanya at Hewlett Packard in the early 1980's

Soon after, Fanya applied for a job at Hewlett-Packard, which is near Stanford and has close ties to the AI Lab there. She was hired as a research scientist in AI at Hewlett-Packard.

"I really loved this job," she said. "I felt like I was on the cutting edge of science.

At Hewlett-Packard Fanya started work on a research program in an area of AI she calls knowledge visualization. From 1983 to 1988 she continued this work at MIT's AI Lab and Media Lab and most recently at Digital's Cambridge Research Lab.

"Imagine a science in which we're looking only at patterns, not the materials the patterns are made of. It's like the study of flight. You can study flight in birds or in airplanes—or even in paper airplanes. It doesn't matter—the principles are still the same.

"So, too, with the study of mind. When you study the mind, you study complicated patterns, and those patterns can exist in brains or in machines."

Building a Smart Machine

"I was always fascinated by Isaac Asimov's idea of a computer that could make another computer just like itself, but only a little smarter. Then that machine, in turn, would know how to build one a little smarter than itself . . . and so on and on, until one would be built to equal or surpass human intelligence. It could happen.

"There's no reason we have to build machines as violent or aggressive as we are. These machines may even be kinder and more peace-loving than we are. War may not make any sense at all to them.

"Doug Hofstadter, an AI person who writes for *Scientific American*, describes an interesting idea (May, 1981, issue). Suppose the military put enough money into AI to build smart bullets, bullets that turn around in mid-flight because they don't want to commit suicide."

When Fanya talks about these interests her face lights up and her hands move excitedly. She has interests outside of her work too. Skiing is one of her hobbies. Music is another interest. Fanya sings medieval ballads in a madrigal singing group. She plays guitar and piano, and sings folk and country songs. Her voice is light and melodic with a bit of country twang.

She lives in a house near the beach with a wonderful view of downtown Boston.

Fanya likes to write poetry. This is a poem that Fanya wrote in October, 1979:

Computer generated portrait of Fanya Montalvo, early 1990's.

SYMBOLS
Rings and groups
circles and FlooPs
generate numerous levels of loops.
Strange,
to be tangled in
Gordion GlooPs.

BlooP

to a space of disorderly soup.

Fanya's poem is about thinking too hard on the same track, about getting stuck in a loop. That is what the Gordian Gloop refers to.

To get out of the rut and get a fresh new idea, may require a leap of imagination. That's what the Bloop is about.

FlooPs, GlooPs and BlooPs refer to specific types of computer programs such as Floo Programs, Gloo Programs and Bloo programs. This is explained in Douglas Hofstadter's book *Godel, Escher, Bach*.

Activities

Trurl and Fibonacci Numbers

Fanya's pet, **Trurl**, is a small crested parrot with lovely gray and yellow plumage.

Trurl is named after the main character in one of Fanya's favorite science fiction books. *Cyberiad*, by Stanislaw Lem, is a collection of short stories about the adventures of Trurl, the Universe's greatest builder of intelligent machines. Such people are called **cyberneticians.**

Trurl can speak a few phrases. He can say "pretty bird." He can say "Trurl is a pretty bird." And, according to Fanya, he whistles very nicely.

Fibonacci is the name of a mathematician who lived in Italy in the 13th century. A number sequence which has many interesting properties is named after him.

This sequence goes on forever. It has no LAST number. Mathematicians call such a sequence an **infinite sequence.**

This is the rule for generating the Fibonacci sequence:

The first number is 1, the next is also 1. From then on, each number is the sum of the two preceding numbers.

?

• **The sequence starts like this: 1, 1, 2, 3, 5, 8, —, —, —, —... Fill in the blanks.**

A The way to do this fill-in is to make your own list of the first 16 Fibonacci Numbers to have at your side. Use this space:

SHADE ALL REGIONS showing FIBONACCI NUMBERS to complete Trurl's picture.

Drawing by Analee Nunan.

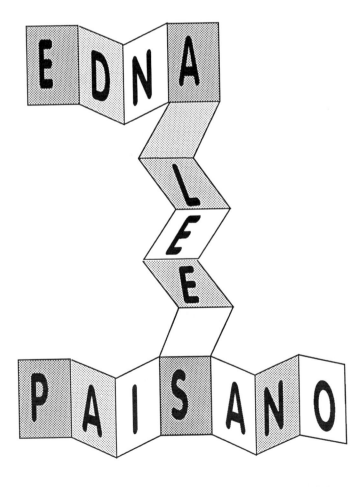

EDNA LEE PAISANO
1948-

**"They taught her to find camas bulbs in the wet Weippe meadow.
She learned how to bake them in a pit in the ground ..."**

Life on the Nez Percé Reservation

Inside the tepee, the girl watched the fire. At the same time, she kept careful watch on the cat to see it didn't get to the meat. Strips of deer, moose, and elk meat hung by the wood fire. As she watched, her mother added more wood to the banked fire and her grandmother turned the cooking meat.

It was the mid-1950's, and Edna Paisano was eight years old. The tepee was set up in the backyard of her family home, near the town of Sweetwater, Idaho. The town is Nez Perce tribal land. By treaty with the United States Government, the Nez Perce tribe owns hunting, fishing, and mineral rights to the land.

Edna is a Nez Perce and Laguna Pueblo American Indian. Her mother, Frances, is Nez Perce, and her father, Ferris, is a Laguna Pueblo who grew up in New Mexico. Edna's parents met in Los Angeles, where many American Indians were relocated by the government for schooling and job training. Soon after they were married, her parents moved to Idaho, to the Nez Perce Reservation, where Edna was born.

Rachel, Edna's grandmother, made moccasins and beaded coin purses from hides to sell. The earnings helped to support the Paisano family. Edna enjoyed helping her grandparents.

A mile down the road from the house were barns and a corral. Fifteen head of the Paisano's Hereford cows grazed on nearby land. The family raised the cows for beef. Their horses grazed near the barns. Two of the horses were gentle enough to ride. After their father came home from work, he walked with the children to the corral, so they could ride the horses and feed the cattle.

The Nez Perces are part of the Shahaptian language family. They were named Nez

Perce by the French, who called them "pierced noses." The tribe was known for its beautiful Appaloosa horses, but by the time Edna was growing up, few Nez Perces owned Appaloosas.

Edna's father worked for Potlatch Forest, Inc., a lumber company in Lewiston near the reservation. He loaded plywood on boxcars for shipping. Her mother at one time worked at the local post office, and in the school cafeteria. But when Edna was in first grade, her mother started back to college and earned a teaching degree. Later, her mother received a masters degree in special education. In 1980, her mother went to Los Angeles to receive the Leo Reano Memorial Award from the National Education Association for leadership in education for American Indian and Alaska Natives.

Learning the Old Ways

Since both of her parents had jobs outside the home, in addition to house chores, animal care, and gardening, Edna spent a lot of time with her grandparents. They taught her how to find camas bulbs in the wet Weippe meadow. She learned how to bake them in a pit in the ground until the whitish bulbs turned dark, and could be eaten or stored throughout the winter. Camas bulbs have a sweet taste, and were a staple food for the family.

Daniel, Edna's grandfather, worked for the State Department of Highways. He helped build a road through the mountains. It was called Lolo Pass, part of the Lewis and Clark Highway. Since Daniel spent four or five days out on the job making the roadbed, Edna and her grandmother went with him and pitched a tepee and tent several miles away on the Lochsa River. There they spent the day fishing, and sometimes they picked huckleberries. When her grandfather returned from work, they ate fresh food for their dinner.

Out in the mountains or back at home, her grandmother told stories and legends to the children before they went to sleep. From her grandparents, Edna learned a little of the Nez Perce language. "Wa-tu" means "no"; "ta'ts maywiy" means "good morning"; "Ne-mee-poo" means "the Nez Perce people."

Edna was very close to her grandparents. She confided in them her growing hopes to gain knowledge about the wider world, and her wish to use this knowledge to help her people.

She says, "My grandparents always encouraged me to do whatever it was I wanted to do. When I was with them, I never had to worry about anything. I just felt so secure."

Edna was dear to her grandparents too. When she was grown up, her grandmother gave her a beaded dress with a design of thunderbolts and triangles.

At home, the family raised a vegetable garden. They planted corn, tomatoes, carrots, radishes and hot peppers. At haying time, she and her brother and sisters helped to haul hay. At branding time, they branded calves. For meat, they ate beef, elk, moose and venison.

Edna Lee Paisano in buckskin dress. Annual Tournament, Nez Percé Reservation, Lapwai, Idaho, circa 1965.

Summer Pasture

The cattle were an important part of the family's livelihood. In the summer, when the temperature in the valley reached 100 degrees, the family took the cattle into the mountains about fifty miles away, where it was cooler and there was good grazing land.

Edna loved summer pasture where her family spent the weekends. The children rode the horses out along the fence line with their father, looking for breaks and helping with repairs. There were views of green meadows and pine trees and elegant red and white cows grazing in the distance. Cool breezes ruffled Edna's short black hair. She and her family slept in tents and cooked pancakes, eggs and meat over an open fire. At night they roasted marshmallows.

Being the Best She Could

When Edna started school in the nearby town of Lapwai, she found out that learning came easily to her. In her school, half of the students were American Indian and half were white. Edna liked being good at her work, but soon she found out that others expected her to be very good all the time. She tried her best to live up to their expectations.

She loved sports and played softball and swam. When she was older, she became coach of a girls' league team in softball. She went to camp, and later became a lifeguard and counselor. Camp was held at state parks and offered swimming, horseback riding, craft and archery. Many of the American Indian children had no chance to be involved in activities the rest of the year.

At school, Edna's favorite subject was mathematics. She always did well in it. In high school, she took algebra, geometry and trigonometry, as well as chemistry and physics. In the math classes, she and her girlfriend had a lot of fun competing with the boys in their class.

"It was she and I against all the guys to see who could get the best grades," she recalls.

But the work was pretty boring for the bright student in the small country school. Only twenty-six students graduated in her class. Edna wished her mind could be stimulated more.

For her first two years of college Edna went to Boise College, which was off-reservation. She took two years of mathematics and again, she enjoyed it. It didn't occur to her to major in mathematics or science, however.

"Could I go back and say to the tribe, `I am a mathematician'?" she asks.

No, she had to choose something that would be useful to her people. At that time, she didn't know how mathematics could be useful. The only choices were farming and teaching. Her education was paid for partly by her family and herself, but also by the Nez Perce tribe and by the Bureau of Indian Affairs, which is part of the United States government. By treaty, this Bureau is supposed to provide money to help educate young American Indians.

Edna wanted to return what was invested in her by working hard at a profession that helped her people. Jobs were few on the reservation. Many people worked at the lumber mill like her father. There were a very few jobs working for the tribal government. Later on, there was a health clinic on the reservation. But, at the time, there was no opportunity for Edna to use her mathematics or science, or to learn what she wanted to know.

Her life had taken a major shift. In order to follow the dreams, she knew she would have to leave the green valley and sparkling rivers and streams she loved, and go outside the reservation.

About this time, Edna's older sister, who had graduated from college and gone off the reservation, took a job in Washington, D.C. She was a teacher for a VISTA project and worked with young Black students. After only a year there, she became ill and died.

The Paisano family felt a tremendous shock. Edna had spent weekends visiting her sister at college. She had looked up to her.

Life off the reservation must have seemed terrifying to the young woman, but a certain way of life was expected of her, and she intended to meet with courage whatever the future offered.

A Disabling Disease _____

One day, when Edna was only twenty years old, she woke up and found she couldn't walk. A disease had struck the active, sports-loving college student with no warning. It was rheumatoid arthritis.

"It was very hard," she recalls. "I used to play softball and swim almost every day, and now even walking is sometimes difficult."

When an attack comes, she has extreme pain in her joints. Over time, the joints break down. When they do, they must be replaced.

In 1974 Edna had her first major surgery. Her hip joint was replaced with plastic and metal parts. After this, she was able to walk more easily. By 1991 however, she needed a second hip joint replacement. This time, thanks to the improvement in medical technology the new replacement allows Edna more flexibility and natural movement than before.

Arthritis continues to be a problem. Over the years, with her doctor's help, Edna has learned to control her disease. Her medication is now down to one pill a day instead of the penicillamine and sixteen aspirin she used before.

She has learned not to walk too far at one time, nor tire herself out. To keep her muscles toned, she swims and exercises.

Moving Off the Reservation

For her last two years of college, Edna transferred to the University of Washington in Seattle. She majored in sociology, graduated and went to graduate school there. Some jobs she held to pay her expenses were: Mt. Rainier park ranger; dormitory student advisor; and recruiter of American Indian high school students for the University. After two years, she earned a masters degree in social work.

Her years at the University were full and satisfying. In addition to her studies and part-time jobs, Edna and some other American Indian students developed contacts in the Seattle American Indian community. They lobbied to have Fort Lawton in Seattle turned into a community cultural center for the American Indians. When meetings and discussions did not persuade the federal government to give them this fort, which was legally theirs, Edna and sixty others held a sit-in at the fort. For this activity she was thrown into jail for a few hours. Eventually, this tactic worked. Fort Lawton now has a thriving cultural center, Daybreak Star Arts Center, for the American Indians living in and around Seattle.

Upon graduating from college, Edna began working at a series of jobs that had some connection with her people and offered the knowledge about new fields she craved. She was a social worker at the Public Health Hospital in Seattle in the rehabilitation ward. Many of her patients were from the reservations.

Moving to Washington, D.C., she worked as a program analyst in the Indian and Migrants Programs Division of the Office of Child Development. In this job, she visited tribal governments in the west and migrant camps in Texas and California. She made suggestions for giving money to these groups, and helped guide Head Start programs for children there.

She worked next for a private consulting firm, helping people in the tribal governments learn rules for good management. She learned how to produce films and slide shows about Indian affairs. She traveled and sharpened her skills in public speaking on reservations, and in Indian centers in cities. Of the six consultants in her firm doing this, she was the only woman.

Because of the arthritis, however, traveling was very tiring for the young woman. She decided to take a job in the government where she would stay in one place, the salary was good, and she had a chance to do something worthwhile.

Working at the Census Bureau _____

In 1976, no American Indian had ever been a full-time employee of the Census Bureau. Edna was the first.

"They had no one to give them our point of view," she says. "I was able to do something new. I was able to provide connections between the Census Bureau and tribal governments and other federal agencies."

While Edna worked on her master's degree, she studied statistics. This laid the groundwork for the kind of work she does.

"It's important for persons who are American Indian and who understand American Indian affairs in this country to take these jobs.

"When I got to the Census Bureau, I realized how important it is for American Indians to know demography, computer programming and statistics: first, because there are very few American Indians in these fields; and second, because the government is always trying to assess things. With American Indian issues, it is very important to have people who can interpret the data accurately."

Every ten years, the U.S. Government Census Bureau counts the number of people and housing units in the United States and gathers information that helps government and private agencies plan for the future. This is important because the number of seats a state is allowed in the House of Representatives is determined by this census count.

Edna was hired by the Census Bureau to work on questions that concerned American Indians and Alaskan Natives. From 1976 to 1980, she and Karen Crook, a co-worker, constructed a special questionnaire that was given to American Indians, Eskimos, and Aleuts who live on reservations, along with the regular 1980 census form. In Oklahoma, American Indians living in areas that used to be reservations were also given the special questionnaire.

The Census Bureau held a "dress rehearsal" testing the special questions on American Indians. From the responses they received, they chose the best questions. They asked about housing, income, education, employment, and health. The answers would help tribal governments make better plans for the future. They would help the United States government live up to its treaty obligations more completely.

During the time of the 1980 census-taking, Edna visited several places to learn how the questions were being answered. American Indians in some locations complained that some of their people had not been counted. If this were true, later, when federal funds were given out to the tribal governments based on the census figures, these tribes would be given less. Edna helped to see that no one was missed.

To prepare for the 1990 census, instead of developing a supplementary questionnaire, Edna's department used more modern techniques to try to get more accurate census data. The bureau focused on ways to inform people about the importance of the census, and how it was in their own best interest to complete the forms. They did this by holding local meetings, displaying posters, disseminating videotapes and sponsoring spot commercials on radio.

Of all of those taken so far, the 1990 census gives the most complete picture of American Indians and Alaskan Natives. Early counts show a 38% increase in numbers of people who are American Indians or Alaskan Natives. People are now pondering the cause of this large increase. Was the increase really this large or was the campaign to encourage cooperation extremely successful?

Edna feels satisfied that it is time to move on to new things, and she has plans for the future.

"I would really like to learn more about estimating, about applying formulas based on certain conditions, and about weighting data. I've done a lot of good in this job, but I need to study more statistics so I can continue to do things which are really important for my people."

Choices for the Future

Going back to school to take these courses is one choice she is considering. Another is to study law.

"Very few American Indians are lawyers and doctors, and there is a great need for them," she says.

But she would not go back to college simply to earn another degree. If marriage occurs along the way, she would be happy.

Edna would only marry an American Indian, but she has not yet met one who suits her needs.

When Edna thinks about the future, and the exciting choices it holds for her, her black eyes sparkle. To learn more, and put her knowledge to use for her people, inspires this woman.

She may decide to live on the Pueblo of Laguna reservation in New Mexico where her father grew up. In the past few years, Edna has visited her relatives there. She has been given gifts of pottery during her visits. She has grown close to these relatives. She could study while she lives there.

"In the end," she says, "I'd like to live back on the reservation—like coming full circle. Going out, then coming back, to share all I've learned."

Activities

WHAT IS STATISTICS?

In the Census Bureau, Edna Paisano works as a statistician. There, people use statistics to draw conclusions from census data.

Statistics helps you make good guesses about answers to situations when you do not have (or cannot have) all the facts or data you need to get an exact answer. Statistics helps you get a pretty good answer from a sample of data.

THE EXAMPLE BELOW USES A STATISTIC CALLED AN AVERAGE

Suppose you want to know how many weeds are in your garden. The weeds are evenly scattered around. It would be tiresome, if not impossible, to count them all. Instead you count the weeds in several small areas around your garden. Each area is a sample. You add the weeds in each sample and get a total. You divide the total by the number of samples you used and get an average. This average is probably a better estimate of the number of weeds in an uncounted section of garden than a pure guess.

THE SECRET CODE ACTIVITY

You can `crack' this secret code by using statistics. To crack the secret code you need to know that certain letters of the alphabet appear very often. Some letters hardly appear at all.

Examine the following experiment. You might use these results, or repeat the experiment and use your own results.

experiment—

Part 1: Get The Data

- Pick a passage from a book, magazine, newspaper or letter.
- Mark a passage that is exactly 100 letters long.
- Make a list of letters of the alphabet like the one here. Go through the passage and make a mark next to the letter on the list, each time you see that letter.

When you get to the 100th letter, total the number of times each letter appeared in that passage. Pick a different passage. Count, mark, and total the letters again.

LETTER	MARK	TOTAL	LETTER	MARK	TOTAL
a	___	___	n	___	___
b	___	___	o	___	___
c	___	___	p	___	___
d	___	___	q	___	___
e	___	___	r	___	___
f	___	___	s	___	___
g	___	___	t	___	___
h	___	___	u	___	___
i	___	___	v	___	___
j	___	___	w	___	___
k	___	___	x	___	___
l	___	___	y	___	___
m	___	___	z	___	___

Part 2: Get Some Samples

The more passages you count, the better your final result will be. It is a good idea to use different sources for different passages; for example, sometimes use a book, sometimes a magazine, sometimes a newspaper. It is a good idea to combine results with others who are doing this experiment. If everyone in a class counts letters in a different passage and results are combined, the final sample will probably be and more accurate.

Part 3: Get An Average

Now you have several sample totals.

You need an average result.

In one experiment, using ten samples, the letters A, B, and C appeared the following times in each sample:

SAMPLE #	1	2	3	4	5	6	7	8	9	10	TOTAL
A	12	13	11	10	10	4	4	7	11	10	87
B	1	2	1	0	1	0	2	1	4	3	15
C	4	2	5	1	1	2	0	3	3	3	24

To calculate an average number of times each letter appears, add the numbers from the separate samples. Then divide by the number of samples. In this experiment, there are 10 numbers from 10 samples. Getting a total for the numbers from each sample, then dividing the total by ten, the following averages were found:

LETTER	AVERAGE		
A	87 ÷ 10	= 8.7	
B	15 ÷ 10	= 1.5	
C	24 ÷ 10	= 2.4	

For each letter of the alphabet, an average was found. Then the letters were arranged in order. The letter having the highest average was put at the top, the one having the lowest average at the bottom. If some letters were tied, any choice of order was made.

This is how it came out:

AVERAGES CHART					
Order	Letter	Average	Order	Letter	Average
1	E	12.5	14	U	2.4
2	T	9.6	15	M	2.3
3	A	8.7	16	G	2.2
4	O	8.3	17	W	2.1
5	N	6.6	18	P	1.6
6	S	6.4	19	B	1.5
7	H	5.6	20	Y	1.3
8	I	5.4	21	V	1.2
9	R	5.3	22	K	0.8
10	D	4.4	23	X	0.2
11	L	3.3	24	J	0.1
12	F	2.9	25	Z	0.1
13	C	2.4	26	Q	0.0

Compare your results with these. Do they look the same? They probably don't look exactly the same. But that's statistics!!

The Secret Message

Now that you know how often certain letters of the alphabet appear, you are ready to crack the secret code. Each letter is in disguise. Use your Averages Chart to establish each letter's true identity. When you substitute the true letter for the disguised letter, the message will appear.

> Dec ika jkair re ma imta re sapesa rvob uabbija. Heviffi
>
> ifs Atoni vatgas Ikora rva pesa ofbrais eq jeofj re rva
>
> Ibgaf ucbop rafr re uaar rvaok qirvak.

Count the number of times each letter appears. Order the letters by how frequently they appear. Compare with the letters in the Averages Chart above.

(**Hint:** The letter that appears most frequently is almost certainly E. The letter that appears next most frequently is almost certainly T; the next A. After that, you may have to do a bit of juggling to decode some of the other letters.)

(**Further hint:** As you decode the message, expect to find the name of at least one person and one place.)

Happy Hunting!

PATTERNS IN A STRIP

Both Edna Paisano and Fanya Montalvo are descendants of people who created beautifully designed pottery, jewelry and fabrics. Edna's ancestors are from two American Indian tribes; Fanya is from Mexico, a country of mixed Spanish and Indian influence. Both women vividly remember beautiful designs on favorite objects they grew up with. Edna speaks about the pattern on a Pueblo bowl she was given in a special ceremony in New Mexico, and of the design on a wonderful dress her grandmother made for her. Fanya speaks fondly of the designs of her people.

Perhaps these women were attracted by the mathematical rhythms of these patterns, as much as by their color and design. Patterns like these, that form a border around a bowl or dish, or a design on a bracelet or belt, are called strip patterns. Few people realize that such patterns are the result of different combinations of slides and flips. In fact, mathematicians who study mathematics of patterns have proven that any strip pattern in the world results from only seven possible combinations of slides and flips.

● **What do we mean by a slide or a flip?**

A Well, **a slide** describes a motion that is a lot like it sounds.

Draw a design; for example

Slide the design some distance and trace it. Now your pattern will look like this:

If you continue to slide and trace, your design will soon look like this:

A A **flip** describes a motion where you turn (reflect) your design before you trace it.

Pretend you were holding a mirror at AB. Depending on where the mirror is placed, the flipped figure will look different. For example, if we label the ends of the mirror A and B, mirrors placed in the following ways would flip the figure the ways you see here.

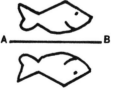

Some patterns are made by combinations of flips. For example, if you flip your original pattern thru line 1, then line 2, then line 3, then line 4, you will have a set of figures that look like this . . .

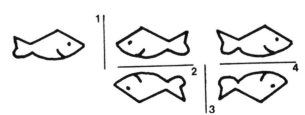

If you erased every other trace, your design would be well on the way to looking like this:

The pictures of the bowls on page 137 were given to Edna Paisano by Pueblo members of her family. They are decorated with strip patterns. Their designs suggest their names... **Bird** and **Lightning**.

Pretend the surface of the bowl is rubber, and that you have cut away the pattern and stretched and flattened it and fastened it to a strip of wood. The figure below shows the basic design unit and its flip transformation for **Bird**.

This is how the pattern is made. The top of the figure below is the design that all the rest is built from. Let's call it the **unit design.**

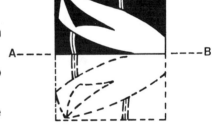

1. **Start by tracing the unit design in rectangle 1.**
2. **Flip the unit design through AB, and trace it in rectangle 2.**
3. **Slide and trace the design from rectangle 2 to rectangle 3.**
4. **Flip the design in rectangle 3 through AB, and trace the design that lands on rectangle 4.**
5. **Slide the design from rectangle 4 to rectangle 5.**

Now you are back in the position equivalent to rectangle 1.
Continue step 2 ... flip and trace.
Then continue step 3 ... slide and trace.
Then continue step 4 ... flip and trace ... and so on.

When you have completed the rectangles in this figure, look again at the pattern on Bird. Do you recognize it now?

1	4	5 = 1
2	3	etc.

A - - - - ... ... - - - B

The unit pattern designs below form the pattern on LIGHTENING. Photocopy the page. These are the design `building blocks.'

Then cut out the unit blocks on this photocopied page, and follow the transformation directions below the blocks to see how to make the strip design.

UNIT PATTERNS

TRANSFORMATIONS

UNIT PATTERN	Place in rectangle 1
FLIP unit pattern through line A, then B	Place in rectangle 2
SLIDE unit pattern right	Place in rectangle 3
FLIP unit pattern through line B, then A	Place in rectangle 4
SLIDE unit pattern right	Place in rectangle 5

Reproduce the Strip Pattern on Bowl B with **unit pattern `blocks'.** See the **transformations** on page 135.

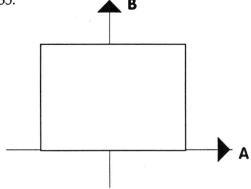

To make the strip design, paste the transformed units on a sheet of paper in the sequence in the figure below.

Continue the pattern.

If you like, copy more unit designs and extend the pattern.

When you are done, check that your strip pattern does indeed look like the pattern on LIGHTNING.

1	4	5=1
2	3	etc.

Now:
- Design your own unit pattern.
- Try different sets of flips and slides to create your own designs.
- Use your designs in weaving, wood carving, leather work, embroidery ... you name it!

Bird

Lightning

THE WISCONSIN THREE

**"If they didn't know the answers
they would not waste my time
by pretending they did."**

The Madison Academic Computing Center (MACC) sits toward the south side of a beautiful college campus, the University of Wisconsin at Madison. Seen from the tower rising behind it, MACC is a flat-roofed white building surrounded by trees. Just a five minute walk north, and the main landmark of the campus, Lake Mendota, is in sight. In winter, the campus is a landscape of bare trees and icy lake. In summer, there is green grass and trees and shrubs—and sailboating on the lake.

Keeping the center running; Left to right: Jean Darling, Sally Handy-Zarnstorff, Kathi Dwelle

The computer center shares a building with the computer science and statistics department. At MACC, some doors are open twenty-four hours a day. Computers sleep only when they "crash," or when they are shut down for maintenance. Otherwise, they are available for use around the clock. In fact, people, or users, are encouraged to use them at "off hours" when rates are cheaper and jobs run faster. This hustle and bustle goes on at the basement level. Here people work at computer terminals and jobs are submitted to the computer. The public offices are on the main floor. On this floor, people can buy manuals and other printed materials that help them work with the facilities. Here, also, consultants answer questions and solve problems people run into when using computers. Consultants are available from 8:30 a.m. to 4:30 p.m. The schedule of consultants is set by MACC, and users talk to whoever is staffing the desk when they need help.

Among the twenty or so people whose job it is to answer questions are three computer programmers: Jean Darling, Sally Handy-Zarnstorff, and Kathi Dwelle. Each woman has taken a different path to reach this point in her career.

"I can remember when I was using the MACC facilities and needed consulting help, I would wait my turn in the line until a consultant was available. When I saw that the consultant was a woman, I felt disappointed. I thought that help from a man would be good; a woman, on the other hand, was not likely to know as much. If the help was not good, it could be very frustrating. Computer runs were expensive, and bad advice could waste time as well as money, and I was always in a rush for results.

*Over the months, these feelings changed completely. After getting help from many different consultants, I found I was very happy to find Jean or Sally or Kathi at the consultant's desk when I needed help. If these women were there, I **knew** I would get my problems solved, or if they didn't know the answers, they would not waste my time and money by pretending they did. Instead, they would quickly get the problem solved by someone who did know. I discovered that women could do first-rate trouble shooting as computer consultants.*

I am embarrassed that my first attitude was sexist, but grateful to them for proving how wrong I was!" —Teri Perl.

JEAN DARLING

1949-

A High School Rebel

Jean Darling grew up in Madison in a family of four children. Her father was a professor in the History of Science Department; her mother was a housewife.

In the early 1960's, Jean was in elementary school, and she loved mathematics. She would work puzzles in the back of *Scientific American* in a column called *Mathematical Games*. Games and puzzles were fun for her. Her father was proud of her and encouraged her. But when she became a teenager, his encouragement stopped. Perhaps he felt that science was not appropriate for a young woman, or that jobs would not be available to a woman scientist. Jean was confused at losing his support.

In high school she began to rebel. She felt that her time was wasted there. Her grades dropped and the situation at school was getting worse. Then, an unusual opportunity came along. She had a chance to study in the north of France for her senior year. She seized this opportunity and left the country.

A year later, when she returned, the high school officials would not accept the French courses for credit. They refused her a diploma, even though she had completed four years of high school work.

Jean refused to let this stop her. She applied and was accepted at the University of Wisconsin at Madison and began her freshman year. She was uncertain about her future. She knew what she didn't want—she didn't want to be a housewife. When she was twelve, she had made up her mind about that.

Trying a "Man's Field"

She wanted to earn a salary as good as a man's, and to do that, she thought, she should go into a "man's field." In 1969, a male friend of hers was earning $5.00 an hour as a computer programmer. This amount would be like $25.00 an hour in 1991.

Jean says, "$5.00 an hour was a lot of money. This guy had long hair and a beard and was real scruffy-looking."

She thought, "He can work when he looks like this and still make a lot of money? This sounds like the life for me!"

About 1970, Jean moved from Madison to Phoenix, Arizona. There, she studied computer science at a technical college and earned a two-year Associate of Arts degree in computer programming. Now she was prepared for a field she liked and one that was well-paid. Very few women were in the field at that time. Now, women make up forty percent of the work force in computer programming.

At first, work was hard to find, but eventually she landed a job at Messenger Corporation, a small printing firm in Phoenix.

In 1971, while she was in school, Jean married Bill Darling. She continued her studies as a programmer. Four years later, in 1975, the Darlings moved back to Madison. There she worked for the State of Wisconsin and for Hopkins Agricultural Chemical Company. By this time, Jean had lots of experience as a programmer, so she became a consultant at MACC.

Jean Darling at MACC, 1980.

She says, "MACC is an unusual shop. It's an excellent place to work. It's not a monstrous organization. You have a lot of independence. People are treated well and they are open to hiring women."

So, Jean's plan is to continue working while taking college courses. She expects to earn her bachelors degree in computer programming. After that she would like to go to Sweden and work for a time.

Community Computers: A Vision

Another goal is to use computers to help community organizations.

She says, "My vision is to have a community database where all the organizations in town make their names, addresses, and phone numbers available. Then a user would pay a quarter or whatever, sit down at a terminal, ask (for example) for all of the anti-nuclear organizations, and get a list of them to contact."

Jean and some other programmers are working toward this kind of community computer in Madison. Work such as this, which is useful to society, is important to Jean. And work helps her to be independent—which she prizes fiercely.

SALLY HANDY-ZARNSTORFF

1953-

"What Do You Want To Be When You Grow Up?"

Sally Handy and her brother, Bill, liked to play together. They both were interested in sports and played them equally well. Often, when people met them and they asked Bill, "Are you going to be a doctor when you grow up?" but asked Sally, "Are you going to be a nurse?" Salley would be very annoyed.

"I didn't want to be rude to this person and say, `Well! you dumb idiot. I have no interest in being a nurse, and just because I'm a girl doesn't mean I can't be a doctor.' I never said that, but that's how I felt. It always bugged me that they thought women could only be nurses and men could be a doctor or anything they wanted," Sally says.

Everyone in her neighborhood played outside. They climbed an old shed roof and played fort in a nearby woods. Neighboring houses had large lawns and no fences separated property. When she ran over the lawns, the owners didn't get angry. They said, "Oh, that's just Sally." The kids organized kickball, softball, and other street games.

At the beginning of junior high school, Sally moved from her small town of Wisconsin Rapids to Madison. She went to the same high school as Jean Darling, but she was a few years younger and the two didn't know each other.

A Short Course in Computers

In her junior year of high school, her math teacher, John Morley, introduced the class to short courses in computer programming. Sally signed up for these courses. The students had a terminal at school and learned to program in FORTRAN and BASIC.

Sally was very good in mathematics and was immediately caught up in her teacher's enthusiasm for computers. Through these classes, she met her future husband, Mike Zarnstorff.

Two Loves: Math and Sports

Sally went away to Carroll College, a small, co-ed, liberal arts college. Out of several hundred in her class, only two women majored in mathematics. Often, in her advanced courses, there were no other women. This never bothered Sally. During one calculus course, two male students came up to her and asked her to tutor them, and Sally did.

Carroll College had women's sports and Sally joined the basketball, volleyball, and track and field teams. "Workouts were three hours a day, five days a week," she recalls.

There was no women's tennis team, but the coach asked her to join the men's tennis team. However, it conflicted with track and field.

Carroll College played teams from other colleges and competed at regional and state-wide conferences. All her friends were in sports too.

Sally says, "Carroll was too small to have the variety of mathematics and science courses I wanted, but on the other hand, it was small enough that I made lots of good friends."

Sally earned her bachelors degree in mathematics. She and Mike were married. She received a master's degree in computer science at the University of Wisconsin at Madison, where the young couple went to live. She began to work toward her doctoral degree, but did not pass the first examinations. She decided to put off further schooling and to begin work in computer programming.

Sally Handy-Zarnstorff at MACC, 1980

Sally was hired at MACC as a consultant. There, she would spend half-time consulting on client problems and was in charge of the other consultants there. The other half, she maintained and developed a tape utility systems program.

Owning a Computer Science Consulting Firm

Sally and her husband and two other friends began their own computer systems consulting firm with clients from industry around Madison. They worked at a terminal in their home which was hooked up to a VAX computer at a research laboratory outside of town. Sally worked at home part-time and would go to the lab three times a week. Working at both jobs occupied her fourteen hours a day.

Working this hard agrees with her. Later, after Mike completes his doctoral degree in physics, Sally plans to work full-time at their consulting firm, or go into industry.

"Computer Science is definitely the career for me," she says.

KATHI DWELLE

1944-

Lighting a Spark

One hot summer day at the end of ninth grade, Kathi Dwelle stopped at school to pick up her final report card. Her algebra teacher called her into the classroom and showed her the results of a standardized mathematics test she had taken.

"Look at this!" he exclaimed. "You missed only three questions out of hundreds on this test. Let's sit down and figure out why you missed these three."

"He really lit a spark in my mind," she recalls.

Kathi lived in a very small town in southwestern Minnesota called Walnut Grove, which was the setting of the television show *Little House on the Prairie*.

In her graduating class, there were only 57 students. Kathi was valedictorian of her class. Her highest grades were in mathematics. She took algebra, geometry, solid geometry, trigonometry and advanced algebra. Her older brother (who now has a doctorate degree in agronomy) used to ask her to help him with some of his tougher math homework problems. By her senior year, she was the only girl in the advanced mathematics class. This didn't bother her a bit; in fact, she liked it.

Kathi's mother was a housewife and her father was a farmer. The nearest university was over one hundred miles away. Her mother had married when she was nineteen and wasn't able to go to college, but she was determined Kathi would get a chance to develop her talents.

Pursuing the Study of Mathematics

Kathi went to Hamline University in St. Paul and majored in mathematics, graduating in 1966. There were no computer sciences departments in universities then, but she thought computer science was an exciting, well-paying field.

She took a battery of tests at IBM and was hired as a junior programmer.

She says, "Industry probably has changed, but in the late sixties at IBM, there was absolutely no freedom. You were very closely watched. There were buzzers to tell you to begin work, have lunch, and quit work. If you arrived after 8:00 in the morning, security guards took down your name and reported you."

"There were dress codes at that time. Males had to wear suits or sports coats and a white or blue shirt. Females were strictly forbidden to wear pants."

Kathi still remembers the day when her manager asked him to help decide if cullotte skirts were acceptable. She voted yes and they were allowed. She also created a sensation when she was the first to wear a mini-skirt to work!

Two years later, Kathi married Dick Dwelle. Life at IBM was not what she had hoped, so she had no regrets when they moved to Madison so that Dick could study for a doctoral degree in English literature.

In Madison, Kathi went to work at MACC as a programming consultant. The young couple liked Madison and Kathi liked her job.

Computer Science—A Good Job

For more than twelve years, Kathi worked at MACC. She started as a consultant, then worked in the graphics department where she learned to design special three-dimensional plotting programs.

Kathi liked graphics design, but after a time she had learned all she could in that department. She moved into a newly created department as coordinator of Contract Programming. (Contract programming means doing programming for users who need special computer help in their work. These users make a contract with a programmer to do this work for some agreed-upon fee.) Kathi supervised two contract programmers and five consultants.

Meanwhile, her husband left the field of English Literature. If he became a professor of English, he reasoned, he might be hired in a small town, where there would be no computer work for Kathi. So he joined a firm in Madison as a stockbroker and found that he liked the work.

Kathi thinks the best part of her work is her contact with people."I work with dozens of highly educated, intelligent people who work in interesting fields. When they come to

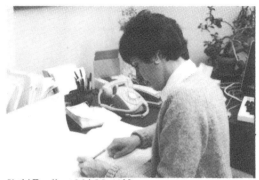

Kathi Dwelle at MACC, 1980.

me to help them design programs, I think, 'What can I do to solve this?' I really like solving problems. I'm always curious to see what the results of their research will be once their data has been analyzed by the computer."

At times, Kathi works fifty to sixty hours a week. She says, "With computers you have to be flexible in your work hours because the computer isn't always working. You arrange your schedule around other things too. Computer time at night is cheaper. If I know a client doesn't have much money, I may come in at night and run jobs at the cheaper night-time rate."

Supporting Other Women

When she first took over the Contract Programming position, she began to deal with clients who were always in a hurry. She worked so hard and felt such strain, her stomach hurt and her hair began to fall out! But soon, she learned to say "No" and "You'll just have to wait!" She persuaded her manager to hire more staff to work with her, and gradually she learned to handle the job pressures.

When Kathi first arrived at MACC in 1968, lots of women were on the staff. Many were earning money to support the family while their husbands were in graduate school. Then, in the early seventies, there was a purge. Half of the staff was laid off. MACC was told it had to "tighten its belt" and balance the budget.

Kathi kept her job but she realized that, by 1976, of a staff of fifty, there were only three other professional women! She and another woman went to the Affirmative Action Officer at the University and reported this fact. They also discussed it with the Director of MACC. As new jobs opened up, the Computing Center began to hire more women. As a result, once again, lots of women work on the staff.

MACC no longer hires and trains people without experience. Only people with programming experience or degrees in computer science are hired today.

"In the early days," Kathi explains, "computer time was very expensive and it was inexpensive to hire and train people. Now just the opposite is true—computer time has come way down in price, but salaries are much higher and so are training costs."

Sharing Equally is Best _____

Kathi and her husband, Dick, decided they did not want to have children. The two of them share the cooking, laundering, and vacuuming equally. They eat breakfast together, which Dick prepares, and dinner, which Kathi prepares. Both of them like to cook.

In Madison, the Dwelles have good friends. Kathi says, "In a university community, you find that your friends are always finishing school and leaving. But now that my husband has a business here, we have friends who are also in business and are staying around."

With Kathi's many interests, she has found that all she really needs is more time. Eventually, she would like to work three or four days a week and have time for gardening, gourmet cooking, canoeing, biking, skiing and travelling.

Beyond the Eighties _____

Ten years have passed since we first wrote about Kathi, Sally, and Jean. We do not know how their life plans turned out.

Did Sally continue her career in computer science? If she did, where did it take her?

Did Jean ever get to build that community database? Has her career helped her to be as independent as she wanted to be?

How about Kathi? Do she and her husband still share household chores? Has Kathi found time for all the activities she talked about? Was she able to build her career as well?

In 1980, all three young women were working at a computer center dominated by one gigantic computer called a main-frame computer. Here users shared telephone lines. The personal computer burst into the world at the beginning of the '80's. How do these women fit into the much larger world of personal computers? Do Kathi or Sally or Jean help people learn how to use personal computers the way they helped people use main-frames before?

We don't know the specific answers to any of these questions. What we do know is that these women were in a very special place early on. Surely their computer science background gave them lots of choices in figuring out how to make their fondest dreams come true.

Activities

MACHINE LANGUAGE

Computers really only understand a language called **machine language.** To run on a computer, languages like **BASIC** or **FORTRAN** or **C** must be translated into machine language for that computer.

Machine language words, on early personal computers, were made up of 8 bits of information. These bits are either 0's or 1's. They are represented, in the machine, as tiny switches that are either on (1) or off (0). Each 8-bit word is called a byte. We can feed decimal numbers into the computer which will then translate them into machine language bytes of 0's and 1's.

PUZZLE

The following activity is a puzzle that uses this idea.
- **Start with a sequence of decimal numbers like 12 or 23 or 57.**
- **Translate these decimal numbers into base 2.**
- **Then write the base 2 number as a pattern of zero's and one's in a row.**
- **Shade all regions where a ONE (1) appears.**
- **Leave all regions where a zero (0) appears blank.**

TRANSLATE DECIMAL NUMBERS INTO BINARY (BASE 2) NUMBERS

To change any decimal number to a binary (base 2) number, write the number as the sum of powers of 2. (Powers of 2 are 1, 2, 4, 8, 16, 32, 64, 128 . . . Notice each number in this sequence is twice the number before it.) The table on the next page compares binary numbers with decimal numbers.

- **Notice: 12=1(eight)+1(four)+0(two)+0(one)** — See line * in table on next page.
- **This is written as 12=1100$_2$**
- **Notice: 23=1(sixteen)+0(eight)+1(four)+1(two)+1(one)**—See line ** in table on next page.
- **This is written as 23=10111$_2$**
- **Notice: 29=1(16)+1(8)+1(4)+0(2)+1(1)** — See line *** in table on next page.
- **This is written as 29=11101$_2$**

	WRITING NUMBERS IN THE DECIMAL SYSTEM			WRITING NUMBERS IN THE BINARY SYSTEM					
	10^2 100's	10^1 10's	10^0 1's	2^5 32's	2^4 16's	2^3 8's	2^2 4's	2^1 2's	2^0 1's
			1						1
			2					1	0
			3					1	1
			4				1	0	0
			5				1	0	1
			6				1	1	0
			7				1	1	1
			8			1	0	0	0
			9			1	0	0	1
		1	0			1	0	1	0
		1	1			1	0	1	1
line *		1	2			1	1	0	0
		1	3			1	1	0	1
		1	4			1	1	1	0
		1	5			1	1	1	1
		1	6		1	0	0	0	0
		1	7		1	0	0	0	1
		1	8		1	0	0	1	0
		1	9		1	0	0	1	1
		2	0		1	0	1	0	0
		2	1		1	0	1	0	1
		2	2		1	0	1	1	0
line **		2	3		1	0	1	1	1
		2	4		1	1	0	0	0
		2	5		1	1	0	0	1
		2	6		1	1	0	1	0
		2	7		1	1	0	1	1
		2	8		1	1	1	0	0
line ***		2	9		1	1	1	0	1
		3	0		1	1	1	1	0
		3	1		1	1	1	1	1
		3	2	1	0	0	0	0	0

In the following example, 62, 34, 2, 14, 14, 2, 34, and 62 are written as sums of powers of 2.

```
62 = 32+ 16+ 8 + 4 + 2
34 = 32          + 2
 2 =               2
14 =         8 + 4 + 2
14 =         8 + 4 + 2
 2 =               2
34 = 32          + 2
62 = 32+ 16+ 8 + 4 + 2
```

In the figure below, a one (1) is placed in each column where that power of 2 is needed to represent the number; zero's (0) are placed where that power of 2 is not needed.

For example:

$62=1(32)+1(16)+1(8)+1(4)+1(2)+0(1)$

Therefore we write $62=111110_2$

Another example:

$34=1(32)+0(16)+0(8)+0(4)+1(2)+0(1)$

Therefore we write $34=100010_2$

This is done for all 8 numbers to complete the grid.

NUMBER	128	64	32	16	8	4	2	1
62			1	1	1	1	1	0
34			1	0	0	0	1	0
2							1	0
14					1	1	1	0
14					1	1	1	0
2							1	0
34			1	0	0	0	1	0
62			1	1	1	1	1	0

Then all squares that have a one (1) inside are shaded.

All those that have zeros (0) are unshaded.

The figure on the left shows the completed picture.

Discover the pictures `hiding' in the grids below.

Remember:

1. Translate each number into its binary form.
2. Record the 1's and 0's in the appropriate columns.
3. Color all squares that have 1's inside.

$28=16+8+4$, therefore $28=11100_2$

NUMBER	128	64	32	16	8	4	2	1
28				1	1	1	0	0
20								
28								
9								
126								
72								
20								
34								

NUMBER	128	64	32	16	8	4	2	1
149								
149								
149								
245								
149								
149								
148								
149								

NUMBER	128	64	32	16	8	4	2	1
56								
108								
56								
145								
210								
86								
52								
24								

NUMBER	128	64	32	16	8	4	2	1
126								
255								
219								
24								
60								

NUMBER	128	64	32	16	8	4	2	1
27								
60								
126								
255								
66								
90								
90								
126								

NUMBER	128	64	32	16	8	4	2	1
48								
56								
60								
62								
63								
160								
255								

NUMBER	128	64	32	16	8	4	2	1
24								
140								
70								
255								
70								
140								
24								

NUMBER	128	64	32	16	8	4	2	1

NUMBER	128	64	32	16	8	4	2	1
			1	1	1	1	1	0
			1	0	0	0	1	0
							1	0
					1	1	1	0
					1	1	1	0
							1	0
			1	0	0	0	1	0
			1	1	1	1	1	1

Make up your own puzzle pictures for a friend.

1. Make a picture by coloring regions black or white.

2. In a blank grid, write a `1' in any square that is black. Write a `0' in any square that is not.

3. Translate these base 2 (binary) numbers into decimal numbers.

4. Give the numbers to a friend. Can your friend reconstruct your picture?

NUMBER	128	64	32	16	8	4	2	1

DECIMAL NUMBERS TO BINARY NUMBERS—SOME SYMMETRIC, SOME NOT

A	B							C
	64	32	16	8	4	2	1	S=symmteric N=non-symmetric
1							1	S
2						1	0	N
3						1	1	S
4					1	0	0	N
5					1	0	1	S
6					1	1	0	N
7					1	1	1	S
8								
9								
10								
11								
12								
13								
14								
15								
16								
17								
18								
19								
20								
•								
•								
•								

Continue the following list:

In COLUMNS B write the binary equivalent for each decimal number in COLUMN A. (See page 149 to learn how to change decimal numbers to binary numbers.)

In COLUMN C write S if the binary number is symmetric. Write N if the binary number is not symmetric.

A pattern is symmetric if it can be folded in the middle such that one half covers the other half exactly.

Since binary numbers are numbers made up of

a pattern of 0's and 1's, we can look at these patterns and decide whether they are symmetric or not.

Using this definition, 3 would be symmetric, since 3 is 111_2 in binary notation, and 1 1 1 is symmetric on both sides of the dotted line.

Similarly, 7 is symmetric since $7 = 111_2$ and 111_2 is the same on both sides of the dotted line.

On the other hand, 4 is NOT symmetric since $4 = 100_2$ and 100_2 is NOT the same on both sides of any vertical dotted line you could possibly draw.

THE UNIVERSITY OF WISCONSIN
Shade all regions containing decimal numbers
that are symmetric when translated into binary form.

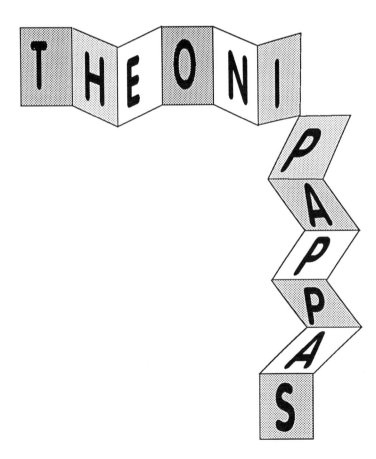

THEONI PAPPAS
1944-

**"... to take things that people see around
and make a math connection ..."**

The yoyo spun and twirled, looped, double looped!   It would do anything she wanted it to do!

Theoni was good in lots of sports. She loved bike riding, hiking, tennis, even scuba diving. And she was good at them.

With her yoyo twirling, she even won championships and prizes at the regional level!

Theoni spent her childhood in Palo Alto, a small town in California alongside Stanford University and near San Francisco.

Palo Alto is a pretty town. You can see mountains on two sides. The mountains on the West hide the Pacific Ocean.

As a young girl, Theoni would often wander over to the family restaurant and watch her father and uncles working. His older brothers were already in the United States when her father came here from Greece as a young man. They were working as waiters in restaurants in San Francisco. In fact, her father's first job was as a busboy at the famous St. Francis Hotel.

In time the brothers decided they knew enough to start their own restaurant, and chose to start it in Palo Alto. They named it *The New Orpheum Cafe*.

The New Orpheum Cafe was located at the foot of University Avenue, at a spot that is still called The Circle. The restaurant remained at the original site until it closed in the early sixties.

There were four children in the Pappas family. First came Theoni's oldest sister, Pearl, then her brother John, then Theoni, and finally her little brother, Nick.

Theoni's father had grown up in a Greek village near Delphi, home of the famous oracle. He was a shepherd until he came to America when he was sixteen years old. Theoni's mother was born in Canada, but her parents too had come from Greece.

Although the Greek community in Palo Alto was small, everyone knew each other. As long as she could remember, Theoni could understand and speak Greek. But she could not read or write it. To make sure their children learned to read and write Greek, her parents got a tutor for them when Theoni was in fourth grade. For the next three years, she and her brothers and sister took Greek lessons once a week.

Theoni loved the outdoors and all kinds of sports, but she also loved going to the library. Theoni did as well in school as outside on the playground. What was special though was that she liked school. This was probably not just because she was smart. Theoni really enjoyed learning!

Theoni's parents emphasized the importance of school and learning. Although they had not attended college, they were always supportive, and they encouraged all their children to go to college.

Theoni can't remember when she did not like math. "From kindergarten on up, whenever math came up in a lesson I'd perk up. Even now, if I'm reading a math book I'll stay up most of the night ." So it was natural that she should major in math in college. Berkeley was not too far away, and that was where she decided to go to college.

It was in college that she got her first taste of teaching. Theoni was a natural! It all happened almost accidentally when she started tutoring some of her college friends. She didn't do it for pay, just to help out.

Theoni was enjoying tutoring so much that she wasn't sure what to do when she graduated. On the one hand she loved studying mathematics. On the other, she loved teaching.

It was difficult to choose between graduate school to study numerical analysis or going into the STEP program at Stanford and learning to be a high school math teacher. The STEP program stands for Stanford Teacher Education Program. Theoni decided to give the STEP program a try.

The STEP Program

At the orientation when she heard ... " In a few weeks you'll be out there with your own class of thirty to thirty-five students," she almost fled on the spot! "Oh NO! This is not for me! I'm not going to do this!" she thought. Luckily she decided to give the summer program a try!

The STEP program was really special. It used unusual methods to prepare people to teach. In the summer program, for example, the student-teachers worked in small groups. They'd prepare short lessons and try them out on local high school students. These teaching sessions would be videotaped. Then the tape would be critiqued and the student-teachers would get lots of pointers.

Theoni started getting less and less nervous, became more and more relaxed. By the end of summer she decided to stay in the STEP program. Then after two more months of orientation it was time to do her internship. Theoni taught geometry and algebra at George Washington High School in San Francisco! In fact, her geometry class had over 40 students!

The class went really well. The STEP people came all year long to videotape and critique her teaching. Her students also got to say how they liked her teaching. All these things together were very helpful.

By 1967 Theoni started looking for her first teaching job. That first job was in Ravenswood High School in East Palo Alto. There she taught all levels from basic math, which is really arithmetic review, to Algebra 2. Theoni remembers it as a special experience.

Over the years she developed different techniques to help put students at ease. "I'd do things like giving kids the option of doing a test over. After all, the idea of tests should be to help people learn, not to punish them."

"Sometimes, I'd start a new class with a joke. I'd say something like 'This is Italian 4.' Then, when the first kid stood up to leave I'd say. 'Wait a minute. This is math ...whatever...' There was usually a big laugh, and everyone was more relaxed, even if it still was a math class.

"Lots of my students who thought they didn't like math decided that maybe they did, after all. I still get letters from the students saying things like "Can you imagine? I'm a math major! Your class really turned me on to math!"

By the beginning of the 70's Ravenswood was probably about 90% Black. It was de facto segregrated. To comply with anti segregation legislation, the district administration decided to make Ravenswood into a special "magnet " school. The idea was to put together a really exciting faculty that would attract and recruit students from other parts of the school district. This started to happen at the same time that Theoni had decided to take a year off to travel .

So Theoni was off to Africa and Spain! When she returned she found she was still wanted at Ravenswood. The place was very different now. As a magnet school it had been turned upside down. There were new teachers. Lots of exciting new programs. For example, kids could take a variety of academic courses as well as such courses as mountaineering, scuba diving, flying, field trips to Mexico and Italy.

After five years the school district decided not continue to support and fund the program. By 1976 Ravenswood, as a magnet school, was closed.

Now all the bussing was on the shoulders of the black students who were moved out to other schools in the district. The faculty was also scattered. Theoni went to Woodside High School where she taught until she left teaching in 1983.

Starting the Business

At Ravenswood, Theoni had met Elvira Monroe. Elvira was teaching English and Journalism when the two first met. Both women were the type of people who got good ideas. And when they got an idea that they thought was great, they were determined to make it happen. That's how their company got started.

Elvira and Phil Arnot, another one of the teachers on the Ravenswood faculty, had just written a book called *Exploring Point Reyes*. The three of them started a company to publish it.

Exploring Point Reyes was the first publication of their new company. It was a real hit, and is still in print many years later. Other books by Elvira and Phil followed.

Then Theoni and Elvira got another idea! How about mathematical ideas on T-shirts? It could be a great way to communicate ideas to students! So that's how *Math T-shirts* got started.

"A teacher can wear a T-shirt that has something on it to make their students curious.

For example, suppose the teacher is going to introduce the Fibonacci sequence or the Pythagorean theorem or the Golden Rectangle.

"He or she would come into class wearing a T-shirt with a picture of a Golden Rectangle on it. The teacher wouldn't say anything, just start the class in the usual way, maybe reviewing homework. You can be sure that sooner or later someone will ask, 'What is that on your shirt?' Then the teacher can launch into the topic illustrated."

First Asilomar Conference

The Asilomar Conference is a meeting of math teachers that takes place once a year. It is always the first weekend in December. The setting is always at Asilomar, a beautiful place on the edge of the Pacific Ocean. It was at Asilomar that Theoni and Elvira were going to see if teachers thought their T-shirt idea was as great as they did.

The two had screened about seventy shirts to take to the meeting. When they got to Asilomar they found the large hall where exhibitors displayed their wares consisting mostly math books and math materials. Theoni and Elvira set up on the half table space they'd reserved. On it they put their shoe boxes with the T-shirts in them.

People came around and discovered the math shirts. By the first day they had sold out everything they'd brought, and were taking orders to fill when they got home. The next day they had a leisurely breakfast and took their time getting down to the exhibit area. "When we finally got there we found people in a long line, waiting for us," Theoni explained. It certainly seemed like there was something important about their product. This one idea was to lead to many, many others.

The Business

One of the next ideas came along when the new company was exhibiting at a book fair in San Francisco. Several books were now in print. *Exploring Point Reyes* was one of them. Another was a book about jogging called *Run For Your Life*. Yet another was *San Francisco— A City to Remember*.

Today Theoni and Elvira's company publishes many different kinds of books by various authors.

"I was walking around during a break and saw a booth that was displaying calendars.

Suddenly I had an idea! Why not a math calendar? Each day could have a problem whose answer would be the date!

"I was so excited, I headed straight back to our booth. 'Hey! I've got this great idea!' " Theoni enthused. Elvira thought it was a great idea too.

"All this happened in October . It was going to be a rush to get the calendar ready for the new year, but we were determined to do it! To save time I wrote all the problems by hand. We printed only a small number of calendars. We wanted to have them for this year's Asilomar conference. Off we went to Asilomar with our three hundred calendars!"

The math calendar was a smash success at the conference.

"At about the same time a friend of my brother's was working at a bookstore near Stanford, and offered to put them in the book store's window. And that's how we started to get the calendars out to the general book market."

It took a lot of work getting the calendars done the first time. "I did the art work myself. I still do much of the graphics for the calendars. Sometimes the art is computer generated. But mostly I like to do it myself so I can be sure that it ends up just the way I want it to be."

All this was going on while Theoni was still teaching. She was incredibly busy. Yet it was a while before she could really let go of her teaching. It happened gradually.

"By 1978 I was teaching three fifths time. To make things even harder my mother became very ill. In a few months she was to die of cancer. She was sixty-five years old. This was very hard on all of us, especially on my father. He had a stroke a year later. His left side was paralyzed, but he managed to rehabilitate himself. About four years later he died from a major stroke. All this was a double shock."

In 1983 Theoni took a two-year leave from teaching to try to catch her breath. When the two years were up she had to decide whether or not to return to teaching. It was a difficult decision, but finally she decided to resign. "I still really love teaching. I often teach math lessons in friends' classes to keep my hand in."

When Theoni taught school, she would do lots of traveling during the summer vacation. She's been to Greece about fourteen times. Visiting Greece is fun for Theoni since she has lots of family there.

"I used to travel almost every year when I was teaching, and lots of times I would go to

Greece, often on the way to, or from, somewhere else." She hadn't gone back for about ten years, until this past fall. Theoni had a marvelous time, even though seeing how Athens had changed was sad. Athens had become very congested and polluted. "When you get out to the countryside you can still recapture some of the spirit and mood of Greece." Theoni got to visit family in both city and country. And this past trip she even went to a couple of islands that she'd never seen before.

Another lifelong passion of Theoni's is her interest in music. As a child she studied the violin. Nowadays she likes to play piano and guitar. "I started playing when I was teaching. I'd always

Theoni enjoying a soft drink while on vacation in Greece, 1991.

wanted to learn flamenco guitar. There's something about music from Latin countries. Even though I had grown up with Greek music, it was somehow different. The first time I heard it, I knew it was for me! It must be something about the right frequency."

Growth of the Business ... the message to get across ... _____

Theoni talks about Math Products Plus and its future. "Our goal is to make people comfortable with mathematics. Math should be something that people aren't afraid of. After all, math is everywhere around us, if we know how to recognize it."

A lot of people are afraid of math. "As a teacher I'd feel it in the air the first day I'd walk into class. Too many people in general are afraid of math. Young people are no different. This shows up in lots of ways. People even feel it's OK **not** to be good in math.

I remember when we first tried to place our calendars in stores. Some buyers, just because they personally didn't like mathematics, said 'Oh NO! Why would we want THAT in here!' They weren't even going to give their customers the chance to decide!

"We want to try to make people comfortable with mathematics. That's why we like to put mathematical ideas on everyday things like bumper stickers and mugs, and those T-shirts we started with. We try to use fun messages such as the one on my favorite math bumper sticker which says 'Start counting ... honk when you reach infinity'. "

Penrose studying his mistress' work.

"We keep trying new things. Every time we come out with a new catalog we include a new kind of product. Last time it was 'post-its' with mathematical messages. We want to take things that people are used to seeing around that are not math connected, and make a connection for them."

Theoni created another innovative approach to mathematics in her book *Math Talk*. *Math Talk* is a collection of poems written for two voices. Two people read the poems out loud. There are certain rules. Each poem is written in two columns. If a line ends on the same row in both columns, both people read the lines at the same time. Otherwise the two voices alternate.

"When I originally saw this type of poetry I wanted to try it with math ideas!" *Math Talk* is filled with poems about all kinds of mathematical objects like circles, the number one, fractals, imaginary numbers, and on and on and on. See the activities section for some poems from *Math Talk*.

What does Theoni want to do next? She is interested in combining mathematics, humor and comics. She has just finished a new book called *Fractals, Googols and Other Mathematical Tales*. It explores mathematical ideas and makes them come alive by relating stories or parables that have appeal for all ages. And of course each year she creates *The Mathematics Calendar*, *The Children's Mathematics Calendar*, and *The Mathematics Engagement Calendar* with all new problems, material and graphics. And each year the number of people who buy the calendars grows.

Theoni and her business partner Elvira are two idealistic women who have built a successful business that supports their beliefs and passes along messages that they want to support. We think they're very very lucky.

Activities

Dinner Anyone?

Theoni and Elvira have written a cookbook called *Greek Cooking for Everyone*. It has lots of family recipes and pictures from their travels in Greece.

Avgolemono soup is a very typical Greek soup. The following recipe gives the correct amounts for a small four-person family dinner.

Suppose you want to try this recipe for yourself alone.

AVGOLEMENO SOUP
8 cups chicken broth
1 cup rice
4 eggs(yolks separated from whites)
3/4 cup lemon juice
2 to 3 stalks of celery, sliced
3 to 4 carrots, sliced

To chicken broth, add rice, carrots, celery and onion. Bring to boil. Cover. Simmer for 20 minutes.

In a bowl, beat egg whites until stiff. Add yolks. Beat well. Slowly add lemon juice, beating continuosly. Without stopping your beating, slowly add 2 cups of the chicken broth. If you stop beating, you risk curdling the soup. When mixed, pour back into remaining broth and rice. Stir over heat. Do not boil. Serve immediately.

???? How would you have to change the amount of each ingredient to make the recipes work for only one person ?

How about one for you and your best friend?

How about a dinner for eight?

What about a big party for twelve?

Make copies of the boxed recipe and complete the blanks for the different size parties.

AVGOLEMENO SOUP
___ cups chicken broth
___ cup rice
___ eggs(yolks separated from whites)
___ cup lemon juice
___ to ___ stalks of celery, sliced
___ carrots, sliced

Poems for Two Voices

Zero

This poem is from Theoni's book *Math Talk*.

Get a friend to read this with you. One person reads the right side and the other reads the left. Lines that appear on the same horizontal are meant to be read simultaneously.

Enjoy!

I am zero. Some say I'm nothing.	I am zero. I have no value.
I know to the contrary.	
	I'm essential
invaluable.	
	I'm the origin on the number line.
The positive numbers are to my right.	
	The negatives to my left.
I'm neither negative	
	nor positive.
I'm zero.	I'm zero.
Centuries before I appeared	
	number writing was burdensome.
repetitious I was discovered	confusing I made the difference in the place value system.
Now with zero there is no mix-up,	
	101 looks different than 11.
Without zero	
	there would be no
place-value system.	
I am zero. Add zero to any number the result is unchanged.	I am zero. Multiply a number by me zero always results.
Divide a number by zero There is no answer.	Beware when dividing by me. The result is undefined.
I am zero. I am nothing.	I am zero. I am essential.

A Story

In Theoni's new book *Fractals, Googols and other Mathematical Tales*, the story, *The Shapes Convention*, describes the sudden appearance of a strange new shape, the fractal. The following excerpt is from that story.

...An argument was brewing about who would lead the opening ceremonies parade. The triangles pointed out they had the least number of sides. But the circles jumped in and said they had no sides and could easily roll out. Finally, a compromise was reached and the triangles and circles were to lead the way together. All the shapes had lined up and were ready to go when they all gasped in unison—"AH! Where did you come from and who let you in this convention," the square demanded. "I'm a fractal?" asked the circle. "I'm sure you have come to the wrong place," the circle continued.

"Oh, no!" said the fractal. "I'm very new—only a few hundred years old. I am one of the shapes of modern times."

"I heard that mathematicians were working with new forms, but had no idea they had materialized one," said the dodecahedron. "Why can't they just leave well enough alone. We were doing a fairly good jog describing the world."

"To the contrary," said the fractal. "There are many things in the universe that your shapes cannot adequately describe."

"An insult," said the square. "How dare you!"

"Not an insult, only a fact," replied the fractal. "Your shapes are fixed and determined. We fractals can grow inward and outward. Just the way things grow in nature. Take me, for instance...

To find out what happens and to read the entire story, check out the book *Fractals, Googols and other Mathematical Tales* from your local library.

Calendar Capers

Try *The Case of the Missing Square* from the month of October of *The Children's Mathematics Calendar 1993.*

"Hello Watson," Penrose said, greeting his old friend. "What are you doing?"

"Oh, I'm trying to figure out a problem," Watson replied. "My mistress has a special 8 by 8 foot square platform made in the following way."

"What do those dotted lines represent?" Penrose asked.

"Oh, that means the square is made up of those four pieces," Watson answered confidently.

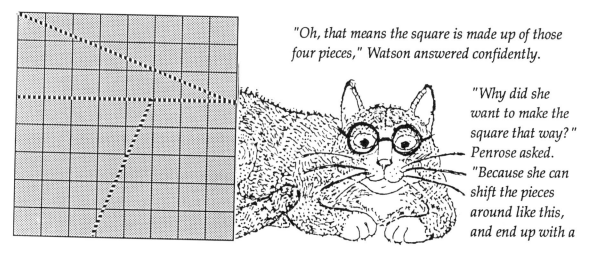

"Why did she want to make the square that way?" Penrose asked. "Because she can shift the pieces around like this, and end up with a

rectangle," Watson demonstrated.

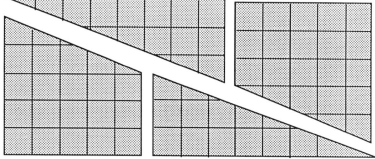

"That's pretty clever," Penrose commented.

"I know," said Watson, "but there is one problem. The square is made up of 8 rows of little squares. That is a total of 64 little squares. While the rectangle is made up of 5 rows of 13 little squares, for a total of 65 little squares. It seems that when the rectangle goes back into square form a little square is lost."

"Wow! This is definitely a problem for my Sherlock Holmes hat," Penrose said as he placed the hat upon his head.

Whenever Penrose wanted to tackle a really difficult logic problem he wore his Sherlock Holmes hat. Somehow the hat added to his confidence and he felt his logic improved. Wearing the hat, he made a model of the platform. He moved the pieces back and forth between the square form and the rectangle form. He took special care to inspect all the details of the problem. Watson watched as Penrose analyzed the problem. Penrose's eyes suddenly opened very wide, and Watson knew Penrose had solved the mystery.

Why don't you make a copy of the square on page 166. Next cut the square along the dotted lines. Study the formation of the square and the rectangle. Can you discover what happened to the little missing square? (Look at the end of the book's solutions section.)

Penrose's Amazing Maze was reproduced from **The Children's Mathematics Calendar 1993** by Theoni Pappas. Copyright © 1992. Reprinted by permission of Wide World Publishing/Tetra, San Carlos, California.

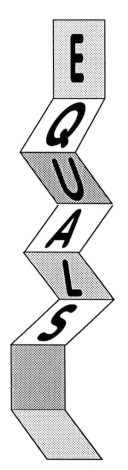

EQUALS
1977-

"She coined the expression — Math! the Critical Filter!"

The Start of Math For Girls and EQUALS

It was 1973. People were saying even as they are now, that girls didn't like math as much as boys. Some girls were turned off because they thought math was not an appropriate subject for girls. And lots of girls were just plain afraid of math! Somehow, it was important to get a different message across. Girls could do math as well as boys! And they certainly should do math if they wanted to have the greatest career choices.

At about this time Lucy Sells, a Berkeley graduate student, was finding that girls were arriving at Berkeley without the math courses required to qualify for many majors. Lack of math was preventing them from taking courses they needed to end up in exciting careers. She coined the expression — "Math! The Critical Filter! "

Nancy Kreinberg first met Lucy at the Lawrence Hall of Science. Lucy was there to get information on percentages of boys and girls registered in the after-school science and math classes at the Hall. As public information person at the Hall one of Nancy's duties was to escort visitors like Lucy.

Sure enough! Three quarters of the classes were filled with boys. Lucy explained to Nancy how math was acting as a filter, working to restrict girls' choices. "Lucy had thrown out the challenge. What were we going to do about it?" "That" says Nancy, "is when three of us started *Math for Girls*."

Rita Levinson was the youngest. She was just starting graduate school and was a math major. Diane Resek already had her PhD in mathematics; in logic in fact. Diane cared a lot about girls getting over fears about math. She wanted all young people to get a glimmer of the excitement that mathematics brought to her, and she had some good ideas about how to do this. Nancy was already working at the Hall, and was a great guinea pig because she herself had lots of fears about math. If an activity sounded like fun to Nancy, it would probably work well with other math-anxious females. With

Diane developing the original course material, the goals were set, the format developed, and the teachers trained.

When *Math For Girls* had been going on for some time, the seed for the next stage began to develop. To make the goals of Math for Girls really work, the program had to spread much faster. "To make it really work we had to work with teachers, with parents, even reach school superintendents." That was the beginning of EQUALS!

Today EQUALS is one of the most important programs for spreading the word about what good math can be all about. It has had an important impact on teachers, by helping them learn how to teach math better. And it has always made a great point that girls, as well as boys, can be successful in math, and that it is important for teachers to help them do that. This story is about the program and several of the people now working at EQUALS.

Heading Up to the Hall

Heading up the hill to talk to several of the EQUALS staff, I drive around the Berkeley football stadium and up Strawberry Canyon Road. The road goes almost straight up, curving sharply from time to time ... past the turnout for the campus Biological Sciences Station ... then past the road running left to the Lawrence Berkeley Laboratory. I drive on and on until I've almost given up, and then, finally, there it is ... the Lawrence Hall of Science!

I park across the road and walk back to the building, past the great pseudo stone whale sitting out front. As I head toward the building I wonder which of the doors that circles the building is the entrance. At the same time I'm drawn toward the low wall to the right of the building.

Looking down over the wall, spread out before me is a spectacular view of the San Francisco Bay. On a clear day you can see Berkeley at your feet, the Bay sparkling behind it, and in the distance you can see the outline of San Francisco and the Golden Gate Bridge. About half the time though you see pretty much nothing. The fog fills the scene; it's as though the building is floating in a dream.

It's hard to believe that anyone can get any work done in such a gorgeous setting. In addition to having some exciting hands-on science exhibits, and some wonderful classes for kids, the Lawrence Hall of Science is the home base of EQUALS.

Deciding whom to interview at EQUALS was difficult, and there wasn't much help from the people there either. Everyone I talked to had a story about who else should be included because they were doing this or that great thing.

I finally ended up interviewing Nancy Kreinburg who is Director of the program and was one of the founding spirits, Virginia Thompson, who first got the idea for Family Math, Kay Gilliland who's now doing a lot of outreach for EQUALS, Sherry Fraser, who's part of a team developing an exciting new high school math curriculum, Jean Stenmark who has written several of the books for the project and is a specialist in new kinds of evaluation, and Terri Belcher, the "new kid on the block" who's helping to strengthen the focus on equity programs with respect to race as well as the original emphasis on gender.

Nancy Kreinberg

Nancy grew up in and around Chicago. The years from three to thirteen were very quiet ones for Nancy, living with her grandparents who were both sick and old. Her divorced mother was leading a seemingly glamorous life working for Esquire, a New York magazine. Nancy did lots of reading in those days. By the time Nancy was thirteen, her mother had remarried, and her half sister was born.

Nancy describes herself as "the most conventional girl anyone could ever meet." She has always felt a special affinity with girls who care about nothing but clothes, dates and friends. Nancy only did "girl" things. All she cared about was her social life.   Like so many young girls at that time, she never expected to do anything except marry when she grew up.   And when you married a man you married his life as well. Nancy proceeded to do just that.

She loved the University of Wisconsin where she went to college and majored in English. Looking back now she is amazed at the things girls took for granted then. When she started school in the mid-50's, girls had curfews. "We never questioned that we had to be in at night at a certain time, but the boys didn't."

It was in college that Nancy met the young man who was to become her husband. Lew was from a conservative, suburban background, very much like her own. But he found suburbia very boring. He was rebellious and the two became politicized together. Nancy remembers her first sit-ins in 1955, and her first picket line in 1957.

According to the formula, when Nancy married this young man, she married a way of life. He was going to be an academic, an historian. So they moved to upstate New York where he started his first teaching job.

After three years at Clarkson College where he was the entire history department, Lew changed his mind about teaching, and in 1963 the couple returned to Chicago. Lew became an organizer in the Black community. Both Nancy and he were heavily involved in the civil rights movement. Nancy worked. The couple had two sons, Jed and Luke.

By 1970 the formula wasn't working right. Nancy and Lew were divorced and Nancy and her children left for California. There she joined a women's group in which she was involved for six years. These people were her first California friends. This kind of women's group was characteristic of the early women's movement. The civil rights and peace movements had been exciting, but both had been highly male dominated. For Nancy, the women's movement was her very own personal political action.

Nancy remembers vividly the first time she and Lucy Sells met and talked. "I remember that I was very interested in psychology when I was in Madison. I switched my major

Nancy Kreinberg

to English because I needed to take math to stay in psychology, and I thought, 'I can't do math.' I never even gave it a moment's thought! Just because I thought I couldn't do math, I had made such an important decision, to give up psychology. Maybe that was the right thing to do, but I definitely did it for the wrong reason.

Talking to Lucy made it all come together for me. The women's movement was important to me. Helping girls overcome obstacles was important to me! Math for Girls and later EQUALS was going to help make it happen! I'd always been looking for a closer connection between my personal life and my work life, and this was it!"

Virginia Thompson

Family Math is Virginia's major project as a member of the EQUALS staff. *Family Math* was her idea. It is one of which she is especially proud! It is now an important program that is helping expand and spread the goals of EQUALS.

Virginia was born in Washington DC, in 1939 and lived in the Washington area through first grade. The second World War had started and Virginia's father was based in Washington working on sonar (underwater sound) at the Bureau of Standards. After the war the family moved to San Diego and remained there until she finished high school.

As a young girl Virginia loved to read. She was also involved in theatre, mostly stage lighting. And Virginia always liked to play with numbers. She can remember picking plums at her granmother's ranch in the Santa Clara valley. She would amuse herself with such thoughts as "How many buckets must I fill to earn two dollars? How many hours must I work to earn three dollars?"

Virginia's brother was four years younger. The two became much closer as they grew older. It was very hard on the family when he was killed in an automobile accident in 1987.

Virginia went to college in California at UC Riverside where she majored in math, but not for any good reason. "Mostly I didn't like to write. I dropped and broke everything in chemistry lab, but I liked math pretty well."

Virginia remembers that her father was very important to her as she grew up. He was a wonderful teacher for his children. He helped her in math all through high school and college. He would never tell her an answer. Instead he would ask questions to help her discover the answer for herself. If she was really stuck he would always say "draw a picture and write down everything you know." This turned out to be a great model for Virginia's own teaching.

Virginia met Kay Gilliland when both attended the first EQUALS summer workshop in 1977. A friend of Nancy's had sent Virginia an application that she received while traveling around the country. The note with it read, "This looks good. Why don't you apply?"

"You had to apply in partners. So I prepared the application, and sent it in with a note telling them I'm applying just in case someone else applies from my school. I never did get a partner but luckily they took me anyway." When the workshop was over Virginia, Kay and a few others were invited to become part-time EQUALS staffers.

Virginia was particularly excited about working with EQUALS because the people there were so supportive and worked together so well. People were always praising each other for their work. Nancy was especially good at that.

"EQUALS was a lot of work, and we had lots of planning to do to prepare the workshops. We learned the hard way how to use our precious time effectively. Any new activity had to be introduced cautiously because it could take up a lot of time. That was because Diane Resek wouldn't let you go on until she had completely solved whatever it was you were presenting, and she was sure she understood it thoroughly."

Family Math started in the early '80's. "Around 1981 we heard from teachers that parents were wanting to help their kids with math but didn't know how to begin. School people didn't have time to do a program for parents. Why didn't EQUALS do one?

Virginia Thompson, right, the mother of Family Math.

It seemed obvious to Virginia that there were thousands of things parents could do. She had worked with parents as teacher-aides in the classroom. She knew they were taking the ideas home to use with their own children. She also realized that many parents were uncomfortable with math themselves. With Family Math, Virginia and the EQUALS staff have shown parents that they can learn more about math, that they can do math, that they can help their own children.

Nancy got the money to start pilot classes. Virginia put together some inexpensive materials. The first class started in January of 1982 with twelve families. The whole thing snowballed very quickly. Before anyone realized what was happening, classes were starting independently, and suddenly it all began to mushroom.

Currently Virginia is full time at EQUALS and spends most of that time managing Family Math. She's also collaborating on a new Family Math book for pre-school that Jean Stenmark is writing with Bob Whitlow, another staff member.

Kay Gilliland

Kay is warm and always friendly. Yet she really gets upset when she talks about how hard it is to change things and how far we still have to go to create an equal society. "Just look at the Lincoln monument! It's positively frightening! Everywhere you read, man this! man that! man! man! man! I think you can't change history, but you can challenge it with some other monuments and some other heroes!"

Kay was born outside London, in England. Her father was English and her mother American. Times were very hard. It was the depth of the depression. Her father, who had left to find work while her mother was pregnant, disappeared, never to return. They never knew what happened to him.

Kay's mother had been born in Kansas. She was a remarkable woman. After teaching school for five years straight out of high school, she earned enough money to go to college. She went to school in Eugene, Oregon where she studied journalism. From there she moved to southern California and became editor of the Hollywood Citizen News. Later on she went to South America where she was editor of the English section of the Panama Times and traveled all over.

In 1927 Kay's mother married her Englishman and off they went to visit England. Kay was born in 1928. After her husband disappeared, Kay's mother scrubbed floors to earn money to return to the United States. Kay 's third birthday took place on the boat returning to the United States.

Kay grew up in Southern California. The depression had hit hard and when Kay's mother returned to the States she tried to work at things that would allow her to be at home with Kay. She organized a play school.

Things eased when she finally got back into quasi-newspaper work as editor for a southern California newspaper. Suddenly she became ill and died. Kay was only fourteen. After her mother's death, Kay was brought up by her aunt. She remembers her mother vividly as a great lady — very outgoing, welcoming, and supportive of all kinds of people.

As a child, Kay wanted to be a carpenter . A carpenter who had come to put a roof on her aunt's house helped Kay learn things. This was really her idea of fun! Kay put up sawhorses in the back yard and proceeded to build a shed. She soon realized that she was not likely to be a carpenter when she grew up. Girls weren't carpenters!

When she was a little older Kay wanted to be a geologist, so she took a geology course in high school and loved it. When she saw the school counselor, he suggested "You might consider micropaleontology."   "No way!" Kay said. "Who wants to be stuck behind a microscope all day." "You can't expect to be a field geologist. No one is going to let a girl go out into the field. Women don't do that!" Kay has since discovered that women, even at that time, did become geologists, but she didn't know it then.

Kay ended up going to a community college in Pasadena and then to Mills College in Oakland for her last two years of college. Since she was planning to teach, she majored in education with a minor in math.

It was as a math teacher that Kay got involved with the California Mathematics Council, and from there she connected with EQUALS.

Kay Gilliland has been involved with EQUALS since its earliest days.

Kay's work with the Math Council was important in several ways. She was general chairperson of the spring math conference for teachers. She got involved with a program that proved to be a natural for helping girls overcome their fears of math. The project was a conference for young people sponsored by the Alameda Contra Costa County Math Educators (AC^3ME) at which students from all over the area did presentations for other students.

"My students presented the toothpick and coins puzzles that I'd been using with my junior high students. These puzzle activities were successes in several ways. They were an excellent way to develop spatial visualization skills, they were a great way for kids to learn to work together, and they helped young people learn to do presentations for others. Doing a presentation for a mixed audience was hard, but with the conference as the incentive they really learned how."

Kay joined the EQUALS staff the second year of the program after having been a participant in the first year's workshops.

Kay is particularly proud of her son Eric who is engaged in a five-year project building her new house. Eric has primary responsibility for all aspects of the job from carpentry to plumbing to electrical to sheet-metal work. The house is set in the Berkeley hills with wonderful views of the city and Bay below.

Jean Stenmark

Jean is the only one in EQUALS who has lived in California all her life. She was born in Davis, California. Her mother and father were born in California as well. Her father had been in the gold mining business before becoming a rancher.

Jean's mother was a teacher before she married. During the depression, when Jean's father was out of work, her mother worked in the canneries.

The most interesting place that Jean ever lived was in the Berryessa valley on land that is now at the bottom of Lake Berryessa. "We were way out in the country. My dad raised turkeys. That was a wonderful life!"

Jean had one sister who was four years older. The girls started school in a one room country schoolhouse with twelve children. All grades were together. It was like being in a big family. "As a first grader I was sitting there listening to what the eighth graders were doing." "We worked on school projects together. One I remember was a great big salt map of California. Everybody worked on it. Eighth grade kids and sixth grade kids did different parts of it."

Jean especially remembers one glamorous teacher, Miss Bronson, who lived with a family in town. "I visited her one day and there she was in the kitchen with her hand under the table and smoke coming out. It became an inspiration to me. I became a smoker." That was back in the days when teachers were carefully scrutinized. For example, in most places young women couldn't get married and still keep their teaching jobs. Jean doesn't smoke anymore. Nowadays most people know that smoking is dangerous, and definitely no longer glamorous.

Life in the one-room schoolhouse ended when Jean was ten and the family moved to Modesto. Because she had been in a non-graded school, the new school people couldn't decide where to place her. Jean was put into sixth grade so if a change were needed she'd move back instead of ahead.

The whole move was very hard. She was expected to march from class to class to the sound of passing-bells. To make matters even worse Jean was very tall for her age. In fact, she was the tallest kid in the school.

Jean stayed in Modesto through Junior College. She went on to finish college at UC Berkeley where she majored in economics. She did very well there.

After graduating from college Jean went to work at the Naval supply center as a clerk. It was war time and her family was still quite poor. Her job at the center was to process requests for aviation supplies. As with so many wartime jobs, Jean was "frozen" into the job. Although she had passed an exam to go to Washington DC as a junior economist, she was not allowed to leave.

Actually, her Navy job proved interesting in many ways. For example she had to decipher requests from mechanics in the field who were using weird names when requesting parts. When she'd look up the name they'd written there was often no record in the catalog for a part with that name. She had to do some serious detective work to figure out what was intended.

Jean learned a lot on that job. Eventually she was sent to a special school in Connecticut where she learned to take apart an airplane engine and put it back together again. She learned to do the same for a propeller, and propellers have thousands of parts.

After the war Jean studied accounting in night school and then became a San Francisco accountant. Eventually she got married and had three children.

As so many women at that time, Jean stayed home with her children when they were little. When she and her husband separated, her children were all in elementary school, first through fifth grades, and she had to get a job. Jean decided that teaching was the thing to do.

"New Math" was the hot subject at the time. It sounded very hard, and Jean hadn't liked math much anyway. She took a summer math course to fulfill her math requirement. Until then she had avoided math courses.

Jean found the new math approach really exciting. It was all about sets, place value stuff and different bases. A lot of the excitement may have been due to her teacher. The teacher would make claims about certain mathematical ideas that would make Jean so angry she would go home and work like crazy to show her teacher was wrong. That turned out to be a great technique. Jean never figured out if her teacher was doing this deliberately to make her students work extra hard.

After Jean finished her course requirements at Cal State Hayward she went to work in Oakland teaching a sixth grade class. In the next years two important things happened to Jean. She was lucky enough to take Miller Math where she first met Kay Gilliland, and she came up against standardized tests and what they meant.

"For me Miller Math was like rockets going off!" This was around the early seventies. It was the tail end of a program that had been funded by the State of California to help teachers learn how to make math special for their students. "I also began, in the classroom, to figure out for myself what standardized tests were all about ... and I set out to do what I could to change them. That's a big part of what I do now at EQUALS."

As time moved on, Jean's career took a series of turns. First she left the classroom to become a resource teacher, then a semi-administrator. The next step would have been to become a principal or vice-principal, but she couldn't bring herself to do that. "I didn't like what I saw as part of the system. Things were so bureaucratic. I didn't want to be a part of it. There was not much connection between what was happening in school and the central bureaucracy."

"Finally I just got tired of fighting the system. I had high blood pressure and knew I couldn't survive that kind of junk. Even if I had to wash dishes, I was going to quit!! "

"My children were now in college. I'd known Kay for a while. We had been together in the Miller Math workshops and had also worked together on the second Conference for Girls. When I mentioned to Kay that I was leaving the district she immediately said that I must come work for EQUALS."

When Jean first came to EQUALS Nancy handed her two stacks of papers and said "Make a book out of this". These were the notes from the Math for Girls classes. Jean formatted what she'd been given, got an artist, and learned a lot along the way.

"For a long time I was one of the workshop presenters. I don't do that much anymore. I don't really enjoy doing workshops. I prefer writing. It's another way to spread the word and it generates income." Jean finds it awfully hard to find time to write, because of all the conferences, meetings and visitors. "I can't write at home the way Nancy does. I seem to need the human contact to be able to write."

Jean Stenmark is still producing books for EQUALS.

"Coming to work here has been a great experience. Miller Math is long gone since its funding didn't last. In math, EQUALS is now just about the only game in town!"

Sherry Fraser _____

Two members of the EQUALS staff are halves of twins. Both Sherry and Terri Belcher have fraternal twin sisters who are their best friends.

Sherry and her twin sister are totally different. Her sister has pitch black hair, very light skin, green eyes. "She used to stand under my chin. We were like Mutt and Jeff. I was the biggest kid in my class. My sister was the smallest. Now we're about the same size." There are also two brothers in the family. One is older than the twins; the other is younger.

The Frasers were a big, happy family, a family of lots of contrasts. "Father was from a fairly wealthy Vancouver family; one of the founding families of Vancouver. My mother came from Spain, from a very poor family. The two met at a dance just before the war."

"We grew up in a very mixed environment. We'd go to my mother's side of the family and all they talked about was sex, politics and religion. We'd go to my father's side of the family where there would be twelve forks in one direction and sixteen spoons in the other direction, and the last thing anyone talked about was sex, politics, or religion."

"My parents had a really good marriage. My father was very positive and outgoing. He'd do things like round up all the kids in the neighborhood and take them to the circus."

Her mother, on the other hand, had been pulled out of school in eighth grade to help support her family. "When we were kids, mother used to tell us that she was going to go back to school and become a math teacher. It never dawned on me that she'd have to go back to high school before she could go to college ... and she wasn't likely to do that!"

"Mother was a homemaker until my father died suddenly of a heart attack when I was sixteen. A couple of years later mother got a job at Xerox.

Sherry finished high school in upstate New York. She went to State University of New York, where she majored in mathematics.

Later Sherry came to California with her math degree. "Even though I had fine credentials, I couldn't get a job. I'd gone to Cornell Aeronautical Labs and applied for a position in the computer department. I was told that I'd scored the highest anyone had ever scored on the computer aptitude test they gave me. And then the fellow said, "Come for a walk with me. I want to show you the department. He walked me around."

"I was twenty-one then. Every secretary was female and in her twenties. Every manager, systems analyst and computer programmer was male, in his thirties or forties, and had on a white shirt and tie. 'Now where do you think you can fit here?' he asked. 'We can't put you with this group. We have to put you with this other group.' "

As a result, "I ended up getting a job teaching!"

"I started teaching cold. I'd never had a course on how to be a teacher. I hadn't been in a high school for years! I had no idea what a lesson plan was. No one bothered to tell me about text books. I thought there weren't any, so I just made up a curriculum." It was weeks before Sherry found out that text books existed and where they were stored.

Later she taught a class called *Slow General Math* . It was assembled in a special way. Everyone picked the two or three worst kids from their class and gave them to Sherry. At the end of the year this "slow" class turned out the only student, a girl, to get a perfect score on the New York State Regents exam! Sherry taught at that school for four years. Then she decided to return to graduate school.

After some graduate school at UCLA and part-time teaching in Los Angeles, Sherry moved to the Bay area and got a job substitute teaching. "An interesting thing about substituting. I was teaching this class one day and I wrote the teacher a note. It was all about how impressed I was with his classroom — how visual it was — student's work all over — projects — interactive — ways for kids to check."

"You don't know who I am?" the teacher asked. "I was superintendent of this district for the last eighteen years and I just quit because I wanted to go back into teaching." The very next day Sherry had a job in the district!

Sherry attended a conference at Mills for adults and it was called *People Who Count*. Nancy Kreinberg was there. "I remember being fascinated by the people and the program."

Sherry was teaching in Novato when a grant to bring a math equity project to the Novato school district was funded. The project goal was to get girls involved in math and science. By this time EQUALS had started and Sherry had been encouraged to attend an EQUALS workshop.

Hired as a half-time person to run the Novato project Sherry worked particularly hard those first two years. "I taught three regular classes in the morning and worked on the K-12 gender-equity project in the afternoon. Meanwhile I connected with EQUALS and started working there more and more. I finally joined the EQUALS staff in its third year.

Sherry Fraser (left) works on new ways to give students math power.

That was EQUALS 3. This is the fifteenth year of EQUALS ... and it's still going strong!"

Now Sherry spends most of her time at EQUALS with an exciting project that has a very long name. Interactive Mathematics Project, or IMP for short. Taking shape at Berkeley High for the last three years, IMP is the setting for an important new high school mathematics curriculum that supports the goals of EQUALS as it helps change the image of mathematics for all young people. Not only is the content innovative but the way students go about learning the content is new. Students of all different abilities and backgrounds get to work together and get to know each other .

IMP builds on a strong peer tutoring program. Students are actively involved in problem solving, in doing mathematics and enjoying it, and in explaining and defending their thinking to each other. Students work in groups that change every two weeks. They share phone numbers and are encouraged to work together on assignments both in class and at home. Students in IMP consider math an interesting thing to do. All this takes place with kids from neighborhoods where many young people are dropping out of school.

Terri Belcher

Most people that know her know that Terri Belcher is a twin. Hardly anyone knows that she was born in Alaska.

She and Toni were born in Anchorage, Alaska in 1955. She lived there until she was fifteen months old, when the family moved to North Dakota. Dad was in the Air Force.

Terri and her family spent about nine years in the midwest. She doesn't remember much about any of it. She was in third grade in Minneapolis. By fifth grade, the family was in Eugene, Oregon. After that they spent six years in Seattle, Washington where the girls went to junior and senior high school. Terri found the city and her schools very interesting.

As a young child Terri liked everyday things. In school, especially in junior high, both she and her sister loved sports and were good athletes. They played softball, tennis, badminton, volleyball and ran track. Terri was a sprinter, her sister a long distance runner.

Terri and her twin sister were the only children in the family. Her twin has always been her best friend. "My sister knows me better than anybody," she declares.

Terri's sister now lives in Minneapolis where she works in marketing, using her creativity and organizational skills. "We have very different professional interests. She's very artistic. I'm not particularly so."

Terri doesn't recall she or her sister having any bad experiences in school. "I don't remember experiencing racism as a little kid or being excluded from clubs or student government. In fact, really the opposite. As twins we were kind of special. As the first grandchildren in our family we were also special." At Mills they were treated personably, as was everyone. Again Toni and Terri played, worked and studied, knowing the world was full of options. In this college for women they were encouraged to try new and non-traditional things. "I learned to raft, ride horses, hike and camp. I went to plays and concerts."

Both parents were important influences on the girls as they were growing up. Their mother has been an important role model. She worked first as a registered nurse, then moved through the ranks to leadership positions in various hospitals. She went on to teach nursing at a junior college. Eventually she moved into administrative jobs. Now she is president of Minneapolis Community College.

Teri's father was in the Air Force for much of his career. Later he was in charge of veteran affairs at Washington State University until he joined the counseling staff at a community college in Seattle. At six feet four, he was an imposing figure. "He always reminded us to 'stand tall'. There was nothing he thought we could not do, even when times got tough."

Terri remembers another important influence in her own life — a wonderful teacher at Mills who made her believe she could do mathematics. Steven Givant had a different way of looking at mathematics, one that made it extremely accessible. Probably because of him she ended up majoring in math.

Before she graduated, Terri took some computer science courses from Carol Lennox. Lennox made learning fun and challenging. She loved teaching and interacting with students. She built confidence in her students by the way she interacted with them and delegated responsibility in her computer labs.

When college was over Terri thought she wanted to go into computer science, so she became a computer programmer with Western Electric. Later she became a manager.

After two and a half years she became dissatisfied with management and decided there must be more to life than that. She decided to become a teacher.

Her first connection with EQUALS took place in college. She remembers, when she was a relatively new math major at Mills, that some kind of Saturday event was going on and math majors were being recruited to help. A woman named Kay was doing a hands-on thing with kids. She was doing toothpick puzzles.   Much later Terri realized that it had been one of the first Expanding Your Horizons conferences, and the toothpick puzzle lady was Kay Gilliland.

The next brush with EQUALS took place some six or seven years later when Terri was already teaching. She had heard about a math conference for teachers at Asilomar and decided to go. It was the first time she was to see Jean Stenmark. There Jean was, on a platform before this huge crowd of people, opening the conference.

In her third year of teaching Terri met a woman who was constantly talking about EQUALS. "At that time I was mostly teaching math lab and working on basic skills. It was frustrating for the kids and it was frustrating for me."

By her fourth year the post-EQUALS teacher finally convinced Terri to go to an EQUALS workshop. "Other than being wonderful, the workshop was an eye-opening experience for me. I learned all kinds of things I had never known before."

"All the EQUALS presenters were wonderful. We listened to ideas, processed ideas, got all involved. At the end of one day Ruth Cossey, one of the presenters, came up to me and said 'I'd like to talk to you.' She said she wanted to come out and see my classes. After she'd seen how I worked with kids in the classroom we began to explore my joining the staff."

"Imagine how thrilled I was to be involved with all these terrific people. I started working on the EQUALS staff in 1989. I'm still one of the new kids on the block."

Terri still dreams about going back to Alaska to check out the icebergs, the Eskimos, whatever else is there. She has a favorite photograph of her father standing before an Alaskan glacier. "My father, smoking a pipe, wearing a short sleeved burgundy shirt, very foxy, with his foot up on a rock, with an enormous glacier in the background. Absolutely beautiful."

Terri Belcher, "one of the new kids on the block," leading the way in new directions.

EQUALS Now

EQUALS sites are popping up all over the country, and even around the world. Of the six original sites, all but one are still functioning. In addition there are thirty-six Family Math sites around the country. Some of these sites offer EQUALS as well as Family Math programs.

Family Math has started in other countries such as New Zealand and Australia. There are budding programs in Sweden, Canada, and Costa Rica. After hearing about Family Math at the last International Congress on Mathematics Education, someone wrote from Poland asking if Family Math might be adopted in their country!

New EQUALS outreach programs help fill the growing demand from teachers who want to attend workshops that are too full to admit them. Basic EQUALS workshops are now being offered in off-site communities that are more accessible to teachers. A cadre of associates or associate staff do these off-site workshops. Terri monitors the nearby sites. Kay works with others that are far away in places such as Alaska or Australia.

EQUALS's Family Math

Family Math is as strong as ever, working with different community based organizations. "We have been able to give them a program with solid content, a little bit of money, the prestige of being connected to the university, and they find the people to do it!" says Virginia Thompson.

The National Urban Coalition took Family Math as a model for their Say Yes to a Youngster's Future program, adding African and other appropriate history to their presentations. People working with Latino kids are organizing Family Math programs in Spanish. The program is getting started in Mexico.

Expanding The Focus

Many at EQUALS agree with Nancy and Jean who would like to see more resources committed to expanding opportunities in math and science for people of color. EQUALS is increasingly reflecting this focus both in programs and in staff.

Terri Belcher and Karen Mayfield, both people of color, are leading the development in this direction. They meet regularly with a committee of teachers called Access and Equity. The goal of this three year old group is to develop leadership among minority teachers and to bring more teachers of color into the California Math Council campaign where issues are addressed like how to improve math learning and how to deal with students of diverse backgrounds.

Until there are more such teachers in the general population, Sherry Fraser believes the best place to tackle the race equity issue is with the kids themselves, and the IMP program does just that. "By working with the students in a non-tracked system, you end up with a group that reflects the general population. You end up with a rich multi-cultural mix."

However this turns out, certain things are clear. EQUALS is stronger than ever and growing. Kay points out that EQUALS has definitely not outlived its usefulness. "Although sometimes we begin to think that everybody must have the message by now, when we look closely we see that this is not so! There is still lots to do."

Activities

Toothpick Puzzles —adapted from the book, *Math For Girls and other problem solvers.**

These are the toothpick puzzles that Kay Gilliland was doing with a group of girls when Terri Belcher saw her for the first time. Get a pile of toothpicks and try them.

These kinds of puzzles can help your spatial visualization and make you a better problem solver. The hardest thing in solving them can be "breaking set". That means, getting out of a rut when you're trying to think of a solution, being loose, and trying something completely different.

TOOTHPICK PUZZLES

1. Use 17 toothpicks to construct this figure.
   a. Remove 5 toothpicks and leave 3 squares.
   b. Remove 6 toothpicks and leave 2 squares.

2. Make this figure with twelve toothpicks.
   a. Remove 4 toothpicks and leave 3 triangles.
   b. Move 4 toothpicks and form 3 triangles.

3. With 9 toothpicks, make this figure.
   a. Remove 2 toothpicks and leave 3 triangles.
   b. Remove 3 toothpicks and leave 1 triangle.
   c. Remove 6 toothpicks and get 1 triangle.
   d. Remove 4 toothpicks and get 2 triangles.
   e. Remove 2 toothpicks and get two triangles.

4. Use 8 toothpicks and a button to form this fish. Move 3 toothpicks and a button to make the fish swim the opposite direction.

You'll probably invent some new puzzles as you spend time with these. Try them on a friend. Sometimes it's fun to see how many different solutions you can come up with for each puzzle.

*From *Math For Girls and other problem solvers*, © 1981. Reprinted courtesy of EQUALS, Lawrence Hall of Science, Berkeley, CA. ©1981.

SPATIAL CREATURES

Study the creatures in the examples below. See if you can identify the characteristics shared by all the creatures in the first row. Make sure that none of the creatures in the second row have those characteristics. Then put a mark on each of the creatures in the third row that share the characteristics of the first row of creatures. List or tell someone the common characteristics that identify all the creatures that fit.

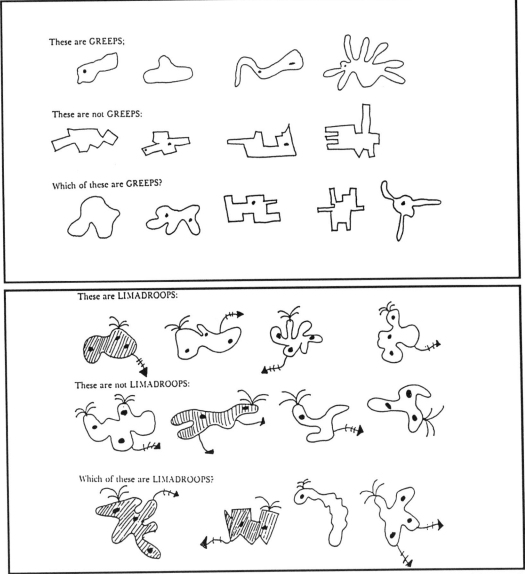

Spatial Creatures—adapted from the book, *Math For Girls and other problem solvers.* Reprinted courtesy of EQUALS, Lawrence Hall of Science, Berkeley, CA. © 1981.

Tax Collector

Why

To practice recognizing and solving problems that involve multiplication factors

The Problem

☐ Your family is about to meet the tax collector! Your goal, of course, is to end the game keeping more money than the tax collector can get from you.

☐ The tax collector **must** however, receive something each time the taxpayer takes a paycheck. Payment is made in the form of **factors** of the taxpayer's check.

▸ *When two numbers are multiplied together, the answer is a* **product.** *The two numbers are* **factors.** *The multiplication problems for 16 are 1×16, 2×8, and 4×4, so the factors of 16 are 1, 2, 4, 8, and 16.*

▸ *A* **prime** *number has only two factors—itself and 1.*

▸ *You may want to review or list the factors of all the numbers before playing the first time (see page 45 for list).*

To Play

☐ Play the first game for practice, using only the numbers 1 through 12.

☐ Put out the twelve paycheck squares across the top of the Tax Collector Board.

☐ Pick a paycheck for the tax**payer,** and put it on that side of the board.

☐ Give the tax collector all the factors of that paycheck (or number).

☐ The tax collector will always get all of the paycheck squares that are factors of the taxpayer's check and that are still available.

☐ Once a number has been used, it may not be used again until the next game.

☐ Since **1** is a factor of every number, the tax collector will get 1 from the first paycheck that is chosen, along with any other factors.

☐ Continue choosing paychecks for the taxpayer and paying the tax collector until there are no paychecks left that have factors.

☐ If there are no factors left for a particular paycheck, the tax collector gets that check.

TOOLS

Paycheck squares, 1 through 24

Tax Collector Board (see page 69)

Pencil

Paper

Tax Collector* is a two-person game. Play it with a friend. You can use the *Tax Collector Board* on page 191.

*From *FAMILY MATH,* © 1986. Reprinted courtesy of EQUALS, Lawrence Hall of Science, Berkeley, CA. ©1981.

☐ When there are no paychecks left that have factors available, the tax collector gets the rest.

☐ Add the taxpayer's total and the tax collector's total to see who has the most.

☐ Here is a sample game:

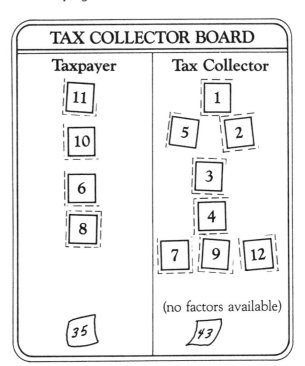

(Well, the tax collector won *that* time!)

☐ Now that you know how to play the game, try it with all 24 of the paychecks. Be sure to work together and talk about why you should choose certain paychecks. Plan ahead. See if you can keep improving your score.

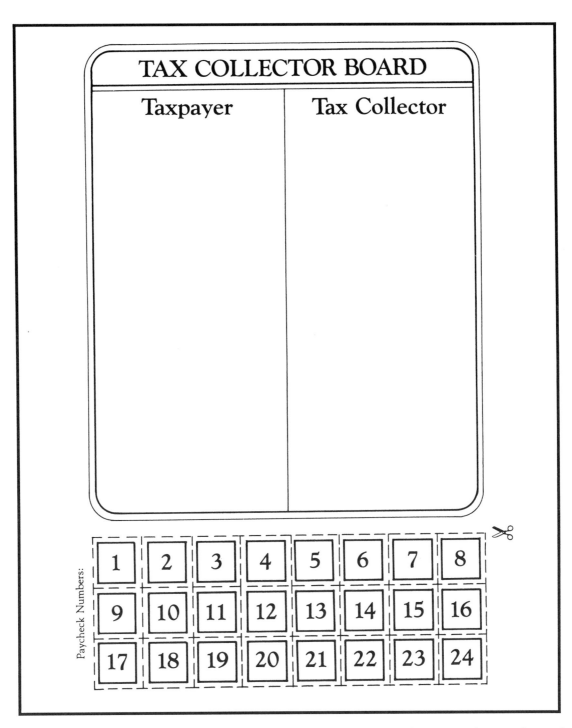

TAX COLLECTOR BOARD

Taxpayer	Tax Collector

Paycheck Numbers:

1	2	3	4	5	6	7	8
9	10	11	12	13	14	15	16
17	18	19	20	21	22	23	24

For an easier game, use only twelve cards to start.... i.e. paycheck squares from 1 through 12. For a harder game, go up to 31 or even 50!

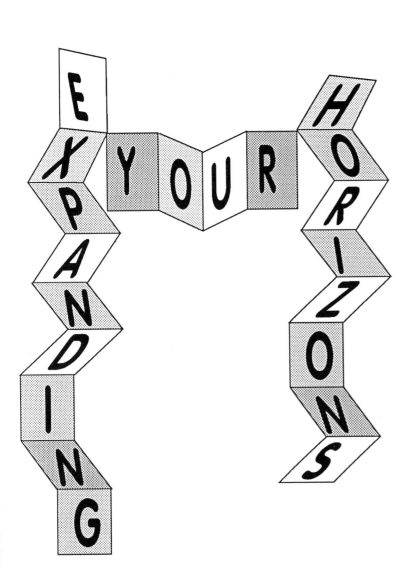

EXPANDING YOUR HORIZONS

by Joanne Pugh

"This conference lets you get guts and go for what you want!"

I glance at the brochure next to me on the car seat. "Expanding Your Horizons in Science and Math . . . A Conference for Seventh to Twelfth Grade Women . . . March 21st . . . Mills College." As I begin looking for the Mills exit along the freeway, scenes from my high school years drift before my eyes. In those days, Miss Farnum, my elderly, grey-haired geometry teacher, was the only female I knew interested in mathematics. My mother hated math and was defiantly proud of repeating Algebra I four times before finally passing. I made a prompt if graceful exit from the bleak landscape of angles and triangles immediately following Miss Farnum's class in tenth grade. The message was forever etched on my mind: women and math don't mix.

Turning onto the campus, I pass Mills Concert Hall. It's only 8:15, but a long line of girls has already formed outside on the wet walkway. As the line moves slowly forward into the Hall, more girls keep coming — alone, in groups, with parents, wearing jeans, kilts, sweatshirts, silk blouses, quilted jackets and slickers. A large bus pulls up near the packed sidewalk. Thirty more girls, all dark-eyed, with black hair, and speaking Spanish spill out onto the street guided by a lone man wearing green sunglasses.

I skirt the crowd and edge toward the door of the Hall. A short, round woman, curlers covered by a vivid orange scarf, shepherds two girls inside and pushes them toward a table marked M-P. I am reminded of early morning crowds waiting to buy tickets to hear their favorite rock group. Today, however, it is science and math, not pop music, which has drawn these young women from their beds so early on a weekend morning. They are only five hundred of the eight thousand girls gathering at that moment on college campuses around California ready to spend the day solving math problems, running computers, and learning about careers in science and engineering.

Inside the Hall, I see Nancy Kreinberg of the conference planning committee. As Director of Math and Science Education for Women at the Lawrence Hall of Science, and Co-Director of the Math/Science Network, she has been working on programs for women and math for seven years. I approach her and ask about the history of the conferences.

"This all began in 1976," she says, motioning to the crowd, "when we held our first conference here at Mills. Two hundred girls attended, and it was very successful. Since then, the demand has been so great that today there are seventeen conferences in California alone. We've spread into ten other states and even overseas. Our newest addition is Australia."

The conference, Kreinberg continues, is sponsored by the Math/Science Network, a group of nine hundred scientists, educators, and community people. They organize events like Expanding Your Horizons to help girls understand why math is important, and to show them that women can be successful in technical fields.

It is 9:30, and the conference is about to begin. For the next hour, there will be a panel discussion on careers for women in science and engineering. The Concert Hall is totally filled, so I sit in the aisle, ready to hear the four women seated on the stage. Two engineers, a veterinarian, and a physical therapist will discuss their jobs, their lives as women in science, their education, and career preparation.

Zella Jackson is tall, slender, and pretty. She is also Manager of Manufacturing Systems at IBM in San Jose and responsible for two hundred million dollars' worth of software products. Somehow, she doesn't fit my concept of a "woman" engineer.

"My algebra instructor probably changed my life," she tells the girls. "She had faith in me and believed I could do it. My father also influenced me by favoring my brothers. I was competing for his attention, so I did things like take shop. When I graduated from Michigan State University in 1974, I was the only woman and the only Black to receive an engineering degree. I love my work. I am challenged and well paid. I also have a viable career with many options."

Jackson's message was one that would be repeated throughout the day: "At your age, don't drop out of anything. Keep your options open and check them all out."

For a moment, Zella Jackson fades away and my mother reappears. I am fourteen years old and telling her that I want to study biology or medicine. She replies, "But you'll have to take calculus and physics, and you won't be able to do it. It's just too hard. Drop out now while your math grades are still good."

Zella Jackson is finished speaking and Dee Jacobson, the veterinarian, has taken the stage. "I recently ran a twenty-six mile marathon," she begins, "and I look at my career in the same way as I regard this race. When I decided to try for the marathon, everyone said, 'You can't do that. You'll die.' I started training and worked at it every day. When the race began, I didn't know what to expect. After twenty miles, I found out. My body ran out of fuel, my legs hurt, I got cramps, but I didn't stop. I decided to be a vet when I was ten. People discouraged me, but I didn't listen. Like the race, I just kept going, and here I am."

By now it is 10:30, and the girls are leaving the Hall to attend math and science workshops. The rest of the morning will be spent in the lab or classroom participating in hands-on problem-solving sessions led by local women scientists, mathematicians, and tradespeople. I take a moment and glance at the material that Nancy Kreinberg has given me.

In elementary school, girls show as much interest and ability in math as boys. But around seventh or eighth grade, these same girls drop math when it is no longer a required subject. During the important junior high school years, female students typically begin avoiding math in favor of English, history, home economics, or "general" courses. Well-meaning teachers and counselors often guide girls away from math and science into courses they view as both more suitable and more useful. These choices, made at age fourteen or fifteen, can have a negative effect on a girl's ability to earn a good living later in life. As our society depends more on computers and technical machines, students without enough mathematics preparation will be left out of many good jobs.

Putting the articles and fact sheets in my notebook, I start out in search of the workshops. The sun is shining now, and the morning has become warm and humid. Wendy Beyea, an operations engineer, leads a group of girls onto the lawn in front of Mills Hall. She is about to begin a surveying exercise, and the girls are carefully following her instructions, adjusting a rod, changing an angle, kneeling, measuring, moving, kneeling, measuring again. Nearby, plumber Naomi Friedman has gathered another group around the back of her truck. Her hands are strong and confident as she carefully sands a piece of copper pipe.

"You need to make sure the pipe is very clean," she is saying, "so the solder will stick." Twelve pairs of eyes follow her motions, as twelve pairs of hands slowly start sanding.

I move into Lucie Stern Hall where classroom doors are open to accommodate the change in temperature. Groups of girls everywhere are solving problems, discovering patterns, predicting answers.

"O.K. girls, check your facts," calls out Dot Mack, an Oakland math teacher working in Room 35. "Good! That was fun: I liked the way you handled the problems."

From another room comes the clicking of dice, as the girls roll and roll again for a probability exercise. Workshop leader Virginia Thompson, mathematics instructor with the Professional Development Program at UC, Berkeley, urges them on.

"What numbers appear most often?" she asks, and the reply, "Sevens and elevens," comes quickly.

Leaving the problem solvers, I join the analyzers over in the Chemistry, Physics, and Mathematics Building. Jacquey Barbar from the Lawrence Hall of Science is supervising an experiment in Room 107. "Tell me," she asks, "What happened when you put those chemicals in the flame?" A young Black woman with thick braids wreathed around her head speaks up in an authoritative voice. As she finishes, Barbar beams, "Good! You're right! You worked really well. I seldom get such a clean set of data and I am very impressed. I wish you all luck." The girls gather their papers and start to leave. It is time for lunch.

Concentration is a powerful key to learning!

As I walk toward Mills Tea Shop, I think about the events of the morning. Never had I seen so many women scientists and engineers gathered in one place nor had I experienced such enthusiasm among young women for technical subjects. When I was young, there were always a few girls, perhaps two or three in a class, who liked math. But I had never seen five hundred together, all so excited about science and math. The effect was very powerful. Today, at the conference, these young women are no longer unusual, different, or painfully obvious. Their interests are confirmed, their career hopes supported, their enthusiasm encouraged.

Comments one ninth grader, "There are only two girls in my advanced-placement math class, and I sometimes feel like an outcast. Now I see other people have done it, and I can, too." I react briefly to her words, feeling some regret about my own past experience, and then pick up my bag lunch from the tables outside the teashop.

Inside, I join a group of people already enjoying roast beef and avocado sandwiches. The conference program includes seminars to inform parents and teachers about a variety of issues important to their children's education. While students are solving math problems and completing experiments, adults meet with representatives of local colleges, attend workshops on career development, and learn strategies to keep their daughters enrolled in math.

I turn to the woman on my left and discover that our children go to the same junior high school. She explains how the morning workshops helped her understand why girls dislike math. We compare notes and find our backgrounds are similar. The difference lies in the gender of our children. Mine, a son, is planning on science as a career. Hers, a daughter, wants to be a writer and is already eager to drop math.

It is nearly 1:00, and the girls are wandering back to the classrooms ready to attend the afternoon career workshops. During the next two hours, students will meet in small groups and form personal contacts with role models—women pursuing careers in science, engineering, and the trades. I glance at the twelve sessions offered and am surprised by the variety of fields. Microbiologist, sanitation engineer, landscape architect, immunologist, biostatistician, medical economist and medical physicist, cytologist, and CPA are just a few of the thirty-six careers represented. This afternoon, together with the young participants, I will learn what a cytologist does, how one becomes a biostatistician, and what courses are necessary to prepare for any scientific field of work. How I wish I had had a similar opportunity back in the ninth grade.

The career sessions are now in progress. I notice that, in addition to requesting specific job information, the girls are concerned about three major issues: family, discrimination, and money. The conflict between family and career is a very real one. The girls don't believe that science and the usual "feminine" interests can be mixed. Questions such as "Do you know about anything else, like art or music?" and "Can you still have children?" indicate the girls' concern about choosing a "masculine" career. I can almost hear a sigh of relief when Diane McEntyre, head of the Mills Math Department, describes how she majored in math, worked for IBM, had her first child, and then pursued studies for a Ph.D. in computer science. McEntyre also notes that "it helps to have a supportive husband."

As long ago as 1980, Expanding Your Horizons conferences were being held in other parts of the country.

The girls are also concerned about being the only woman among all male workers on a job.

"When the men start bothering you," one tenth grader asks, "how do you turn them off?"

Laughter greets this question and also the response, when electrician Diane Kutchins replies, "It's like when the boys tease you in school. You just ignore them."

Expanding Your Horizons conference. in conjunction with Society of Women Engineers (SWE). Madison, Wisconsin, 1980.

The consensus among the role models is that although women earn less and advance more slowly than men, a truly competent person eventually succeeds. Perhaps the most surprising to me, who had always thought self-fulfillment the primary reward of work, is the persistent "How much money do you earn?" These young women are concerned about the economic realities of their future, as well as with fulfilling their personal interests. The salaries quoted are impressive ... and seem enough to encourage even the most math-shy student not to drop out.

It is three o'clock, and the conference is over. As I head back to my car, I am surrounded by a group of girls leaving Lucie Stern Hall. Their excited conversation is easy to overhear.

"I want a career in space sciences," one says to another, "and now I know that taking physics is important."

Her companion nods and replies, "This conference lets you get guts and go for what you want. I'm going to be a doctor, and am definitely looking forward to my career. I know I can do it."

* * *

The Expanding Your Horizons conferences are now in their seventeenth year. They are

Site of the New Math/Science Network office. This is one of the 16 restored Victorian houses at Preservation Park in Oakland, California.

still organized very much as described above. In 1992 almost twenty-six thousand students attended Expanding Your Horizons conferences around the country. There are presently over a hundred conferences nationwide.

Because the national focus on environmental issues has created career areas that were not common ten to twenty years ago, the Math/Science Network created a new one time conference which took place at Mills College in 1990. It was called Expanding Your Horizons in Science and Mathematics for the Environment (EYHE).

Two sets of hands-on workshops provided students with a wonderful exposure to careers in environmental science. The workshops included Sherlock Holmes of Hazardous Waste (unraveling the mysteries of unknown hazardous waste), Keeping Dirt Clean (contaminant properties in soils), Rock and Roll (seismic studies and the impact on waste disposal), Acid Rain (its presence and impact), and many more. The focal point of the day was a "Mock Spill" where students were involved in the effects, clean-up, reporting, and legal ramifications of a simulated spill of hazardous material.

The pictures here show the EYHE conference in action.

Courtesy of The Math/Science Network.

Activities

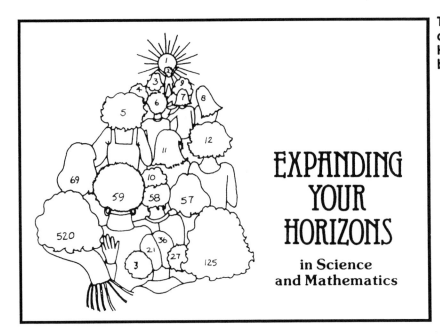

This drawing is from an Expanding Your Horizons conference brochure

When 4 is divided by 3 the remainder is 1. The same is true for 1, 7 and 10.

When 5 is divided by 3 the remainder is 2.

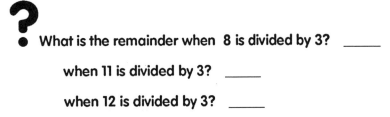

?

● **What is the remainder when 8 is divided by 3?** _____

when 11 is divided by 3? _____

when 12 is divided by 3? _____

To color the picture from the Conference brochure, divide each number by 3 and find the remainder.

If the remainder is 0, color the hair red.

If the remainder is 1, color the hair yellow.

If the remainder is 2, color the hair brown.

So 4 is colored yellow. What is 5 colored?

Solutions are given at the end of book in the section SOLUTIONS TO ACTIVITIES.

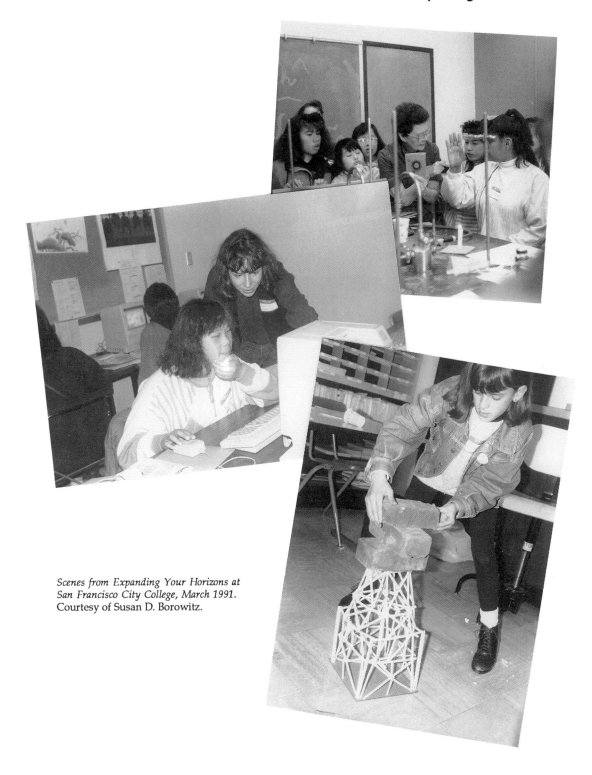

Scenes from Expanding Your Horizons at San Francisco City College, March 1991. Courtesy of Susan D. Borowitz.

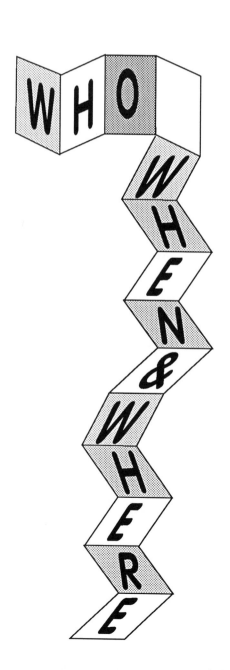

WHO, WHEN & WHERE

Who	When	Where
Mary Somerville	1780-1872	Scotland
Ada Lovelace	1815-1852	England
Sonya Kovalevsky	1850-1891	Russia
Mary Everett Boole	1832-1916	England
Emmy Noether	1882-1935	Germany
Lenore Blum	1942-	United States
Evelyn Boyd Granville	1924-	United States
Fanya S. Montalvo	1947-	Mexico
Edna Lee Paisano	1948-	United States
The Wisconsin Three		
Jean Darling	1949-	United States
Sally Handy-Zarnstorff	1953-	United States
Kathi Dwelle	1944-	United States
Theoni Pappas	1944-	United States
Equals	1977-	United States

Solutions for Activities

page 59
Mary Everett Boole —Drawing by Analee Nunan

page 10—
Mary Somerville —Drawing by Analee Nunan

page 13—
On The Way To The Ball —Drawing by Analee Nunan

page 27—
Ada Lovelace —Drawing by Analee Nunan

page 26

Triangular numbers from 1 to 210:

1, 3, 6, 10, 15, 21, 28, 36, 45, 55, 66, 78, 91, 105, 120, 136, 153, 171, 190, 210

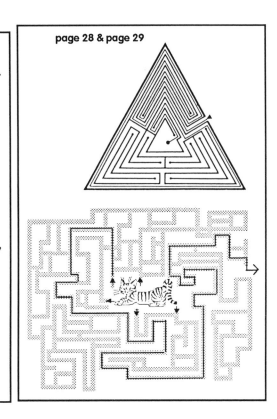

page 28 & page 29

page 42—
Sofia Kovalevsky —Drawing by Analee Nunan

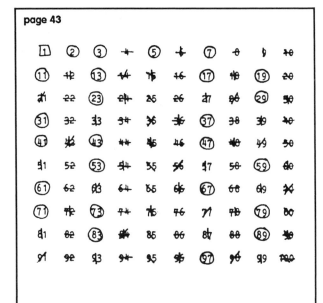

page 43

page 71

5 (+) 8 = 1
8 (+) 5 = 1
11 (+) 4 = 3

page 58

For the first pattern of square dots, each new square of dots is created by adding the next odd number of dots to the previous square's dots.

For the second pattern of square dots, each square of dots is composed of the sum of two consecutive triangular numbers.

For page 59's solution see page 204.

page 92

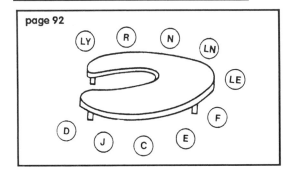

page 75—
Emmy Noether —Drawing by Analee Nunan

page 93—
Caracas —Drawing by Analee Nunan

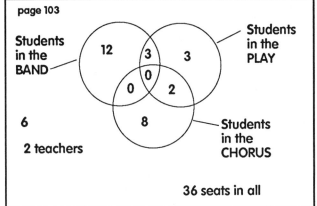

page 103

Students
in the
BAND

12   3   3

0   0

0   2

6

2 teachers

8

Students
in the
PLAY

Students
in the
CHORUS

36 seats in all

page 106
<u>1</u> 4 heads
<u>4</u> 3 heads, 1 tail
<u>6</u> 2 heads, 2 tails
<u>4</u> 1 head, 3 tails
<u>1</u> 4 tails

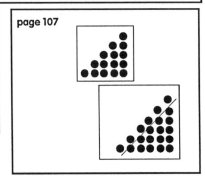

page 107

page 107

A row	B sum across rows	C multiples of 2s	D powers of 2
1	1	1	2^0
2	2	2=2	2^1
3	4	2x2=4	2^2
4	8	2x2x2=8	2^3
5	<u>16</u>	<u>2x2x2x2=16</u>	2^4
6	<u>32</u>	<u>2x2x2x2x2=32</u>	<u>2^5</u>
7	<u>64</u>	<u>2x2x2x2x2x2=64</u>	<u>2^6</u>
8	<u>128</u>	<u>2x2x2x2x2x2x2=128</u>	<u>2^7</u>
9	<u>256</u>	<u>2x2x2x2x2x2x2x2=256</u>	<u>2^8</u>
10	<u>512</u>	<u>2x2x2x2x2x2x2x2x2=512</u>	<u>2^9</u>

page 119

Fibonacci numbers from 1 to 1000:
1, 2, 3, 5, 8, 13, 21, 34, 55, 89, 144, 233, 377, 610, 987

page 119—
Fanya's pet trurl —Drawing by Analee Nunan

page 132

You are great to be able to decode this message. Johanna and Elize helped write the code instead of going to the Aspen music tent to meet their father,

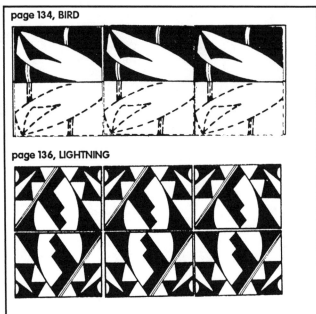

page 134, BIRD

page 136, LIGHTNING

page 151

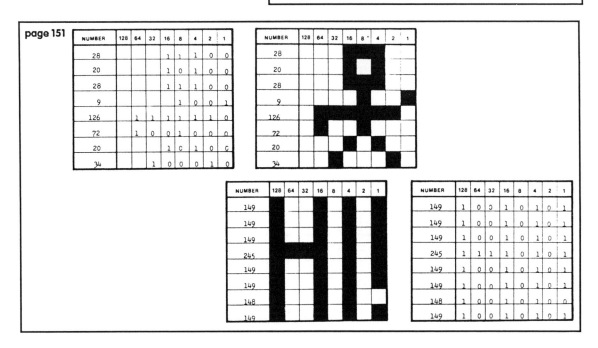

NUMBER	128	64	32	16	8	4	2	1
28				1	1	1	0	0
20				1	0	1	0	0
28				1	1	1	0	0
9					1	0	0	1
126		1	1	1	1	1	1	0
72		1	0	0	1	0	0	0
20				1	0	1	0	0
34			1	0	0	0	1	0

NUMBER	128	64	32	16	8	4	2	1
149	1	0	0	1	0	1	0	1
149	1	0	0	1	0	1	0	1
149	1	0	0	1	0	1	0	1
245	1	1	1	1	0	1	0	1
149	1	0	0	1	0	1	0	1
149	1	0	0	1	0	1	0	1
148	1	0	0	1	0	1	0	0
149	1	0	0	1	0	1	0	1

page 151 continued

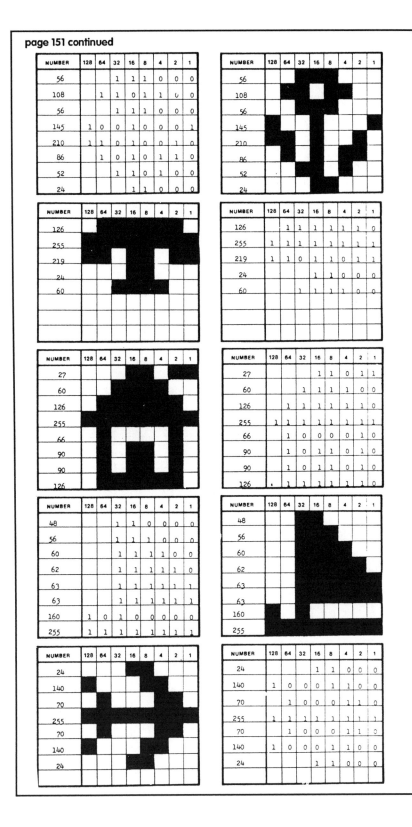

NUMBER	128	64	32	16	8	4	2	1
56			1	1	1	0	0	0
108		1	1	0	1	1	0	0
56			1	1	1	0	0	0
145	1	0	0	1	0	0	0	1
210	1	1	0	1	0	0	1	0
86		1	0	1	0	1	1	0
52			1	1	0	1	0	0
24				1	1	0	0	0

NUMBER	128	64	32	16	8	4	2	1
126		1	1	1	1	1	1	0
255	1	1	1	1	1	1	1	1
219	1	1	0	1	1	0	1	1
24				1	1	0	0	0
60			1	1	1	1	0	0

NUMBER	128	64	32	16	8	4	2	1
27				1	1	0	1	1
60			1	1	1	1	0	0
126		1	1	1	1	1	1	0
255	1	1	1	1	1	1	1	1
66		1	0	0	0	0	1	0
90		1	0	1	1	0	1	0
90		1	0	1	1	0	1	0
126		1	1	1	1	1	1	0

NUMBER	128	64	32	16	8	4	2	1
48			1	1	0	0	0	0
56			1	1	1	0	0	0
60			1	1	1	1	0	0
62			1	1	1	1	1	0
63			1	1	1	1	1	1
63			1	1	1	1	1	1
160	1	0	1	0	0	0	0	0
255	1	1	1	1	1	1	1	1

NUMBER	128	64	32	16	8	4	2	1
24			1	1	0	0	0
140	1	0	0	0	1	1	0	0
70		1	0	0	0	1	1	0
255	1	1	1	1	1	1	1	1
70		1	0	0	0	1	1	0
140	1	0	0	0	1	1	0	0
24			1	1	0	0	0

page 152—

Symmetric Binary Numbers between 1 and 80

decimal	binary
1	1
3	11
5	101
7	111
9	1001
15	1111
17	10001
21	10101
27	11011
31	11111
33	100001
45	101101
51	110011
63	111111
65	1000001
73	1001001

page 153—
University of Wisconsin —Drawing by Analee Nunan

page 163— for 1 person
AVGOLEMENO SOUP
__2__ cups chicken broth

1/4 cup rice

__1__ eggs(yolks separated from whites)

3/16 cup lemon juice

1/2 to 3/4 stalks of celery, sliced

3/4 to 1 carrots, sliced

page 163— for 2 people
AVGOLEMENO SOUP
__4__ cups chicken broth

1/2 cup rice

__2__ eggs(yolks separated from whites)

3/8 cup lemon juice

1 to 1 1/2 stalks of celery, sliced

1 1/2 to 2 carrots, sliced

page 163— for 12 people
AVGOLEMENO SOUP
__24__ cups chicken broth

__3__ cup rice

__12__ eggs(yolks separated from whites)

2 1/4 cup lemon juice

6 to 9 stalks of celery, sliced

9 to 12 carrots, sliced

page 163— for 8 people
AVGOLEMENO SOUP
__16__ cups chicken broth

__2__ cup rice

__8__ eggs(yolks separated from whites)

1 1/2 cup lemon juice

4 to 6 stalks of celery, sliced

6 to 8 carrots, sliced

page 167

When the square is cut into four pieces and made into the rectangle, you will notice a thin gap. This gap totals the area of a little square. It is the missing square.

page 187— There is an "X" on each toothpick to be removed. In most cases there are several possible solutions. Only one is given.

1. a. b.

2. a. b.

3. a. b. c.

   d. e.

4.

page 200—

page 188—

The CREEPS are circled.

The LIMADROOPS are circled.

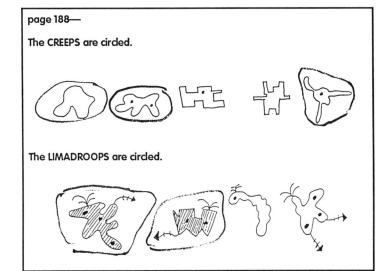

Photograph by Joseph Perl.

Dr. Teri Perl is an educational consultant, software designer, and developer of curriculum materials. She is currently head of a Task Force on Education & Technology for The Association of Computing Machinery (ACM).

Women and Numbers is Perl's second book on women and mathematics. Her earlier book, *Math Equals*, was cited by the American Library Association for its outstanding contribution to mathematics.

Perl has a B.A. in Economics from Brooklyn College, an M.S. from San Jose State University, and a Ph.D. in Mathematics Education from Stanford University.

ENDORSEMENTS

"This book is a must for the reference library of every teacher in grades 4 through high school who is looking for innovative ways to motivate his or her female students to continue their study of mathematics"
— **The Mathematics Teacher**
Mona Fabricant, Queensboro Community College, NY

"Both girls and boys may well read WOMEN & NUMBERS from cover to cover, and adults will be interested as well. I recommend it to all for pleasure and enlightenment." — **Math/science Network Newsletter**
Leonard Feldman, San Jose University, CA

"WOMEN & NUMBERS is a must read! The stories about women and mathematics make them come alive to the reader. The activities are engaging and mathematically meaningful."
— **Lyn Taylor, PhD. University of Colorado-Denver**
Past President, Women & Mathematics Education (WME)

"…attractively-produced, well-illustrated, and challenging…should help many young women to feel that they can do it in mathematics."
— **Claudia Zaslavsky, author & educator**

"The forty 12-year olds in our summer math/physics program for girls devoured the book. It tied interests of these adolescent (or pre-teens) girls, such as clothing, picture coloring, and growing-up stories in various times and cultures to the problem-solving and mathematical concepts created by past and contemporary women mathematicians. It made a mathematics career seem possible and positive for women." — **Al Saperstein, Professor**
Department of Physics, Wayne State University

"Lively, revealing and educational." — **The Children's Bookwatch**
The Midwest Book Review